Practical Career Advice
for Engineers

Practical Career Advice for Engineers

Personal Letters from an Experienced Engineer to Students and New Engineers

Radovan Zdero

CRC Press
Taylor & Francis Group
Boca Raton London New York

CRC Press is an imprint of the
Taylor & Francis Group, an **informa** business

First edition published 2022
by CRC Press
6000 Broken Sound Parkway NW, Suite 300, Boca Raton, FL 33487-2742

and by CRC Press
2 Park Square, Milton Park, Abingdon, Oxon, OX14 4RN

Library of Congress Cataloging-in-Publication Data

Names: Zdero, Radovan, author.
Title: Practical career advice for engineers / Radovan Zdero.
Description: First edition. | Boca Raton, FL : CRC Press, 2022. | Includes bibliographical references and index.
Identifiers: LCCN 2021011469 (print) | LCCN 2021011470 (ebook) | ISBN 9781032044132 (hbk) | ISBN 9781032044118 (pbk) | ISBN 9781003193081 (ebk)
Subjects: LCSH: Engineering--Vocational guidance.
Classification: LCC TA157 .Z43 2022 (print) | LCC TA157 (ebook) | DDC 620.0023–dc23
LC record available at https://lccn.loc.gov/2021011469
LC ebook record available at https://lccn.loc.gov/2021011470

ISBN: 978-1-032-04413-2 (hbk)
ISBN: 978-1-032-04411-8 (pbk)
ISBN: 978-1-003-19308-1 (ebk)

DOI: 10.1201/9781003193081

Typeset in Times
by MPS Limited, Dehradun

"The progressive development of man is vitally dependent on invention. It is the most important product of his creative brain. Its ultimate purpose is the complete mastery of mind over the material world, the harnessing of the forces of nature to human needs."

Nikola Tesla (1856–1943)
Electro-Mechanical Engineer
My Inventions: The Autobiography of Nikola Tesla

Contents

PART I Basics

PART II Preparation

PART III Principles

PART IV Workforce

PART V Aftermath

Acknowledgments

I would like to extend a huge thanks to the entire team at CRC Press for all of their wonderful assistance in bringing this book into being. I appreciate the valuable input and feedback on various portions of the book from my engineering colleagues and other friends: Mr. Aaron Gee; Mr. Aleksandar Djuricic; Dr. Aleksandar Jeremic; Mr. Alex Matheson; Ms. Amy Posch; Mr. Bruce Nicayenzi; Mr. Eric Velasquez; Mr. Geoff Leung; Dr. Habiba Bougherara; Dr. Jelica Zdero; Dr. Mark-John Bruwer; Dr. Shaghayegh Bagheri; and Mr. Suraj Shah. I want to also mention several fine folks who have mentored me in my engineering career: Dr. Özden F. Turan; Dr. J. Timothy Bryant; and Dr. Emil H. Schemitsch. I want to express my immense gratitude to my family and friends who support me in many ways and with whom I experience this thing called life. I especially cannot overstate the debt I owe to my dear parents Obrad and Ana Zdero (a.k.a. Обрад и Ана Ждеро) who taught me valuable lessons about character and competence. Finally, I'd like to recognize the Grand Designer who engineered the cosmos.

Author

Radovan Zdero, PhD, CEng, MIMechE, has decades of experience as an engineer and a mentor to engineers. His engineering background includes a master's degree in aerodynamics (McMaster University, Canada) and a doctoral degree in biomechanics (Queen's University, Canada). He is a Chartered Engineer, a Member of the Institution of Mechanical Engineers, and a Professor in the Division of Orthopaedic Surgery and the Department of Mechanical and Materials Engineering (Western University, Canada). He has published many scholarly research articles in peer-reviewed engineering, science, and medical journals. He is also the editor of the engineering textbook *Experimental Methods in Orthopaedic Biomechanics.* Contact the author: dr.zdero@hotmail.com.

Preface

I'm a professor and senior researcher in mechanical engineering with a specialty in biomechanics, biodevices, and biomaterials. Some people would call me a biomedical engineer. Either way, I think I'd be equally happy being an engineer who designs, builds, and fixes robots, bridges, or spaceships. But, where did this interest in engineering come from? As I look back, I can see the early roots of this passion in my family's influence (i.e. nurture) and my own personality (i.e. nature).

My grandfather was an illiterate peasant farmer in a European village who had a knack for therapeutically massaging and manipulating the painful, sprained, bruised, or broken limbs and joints of other farmers who were injured during work. This is one of my earliest memories in life. My parents were also mechanically inclined. Father was a peasant farmer, factory laborer, and master carpenter, while mother was an expert seamstress also skilled at baking, cooking, crocheting, embroidering, and knitting.

When I was a very young kid, I often disobeyed my parents by sneaking into the tool shed to play with all those interesting tools. For me, playing meant hammering nails into the wooden floor, pushing manual lawnmowers around, and taking out the car jack to lift up the back of my father's car. I sustained at least one injury that remains visible to this day.

Later on, when I was a student in high school, I really enjoyed and excelled in my courses in mechanical drawing, mathematics, and chemistry, but I also liked my courses in physics, biology, and computers. My chemistry teacher used to leave the classroom door unlocked, so my friend and I could come in by ourselves after school hours to do chemistry experiments—unsupervised! It's lucky we didn't blow up ourselves or the school! I even thought about becoming an architect or a chemist.

And so, when it was time for me to apply to university, I wasn't necessarily sure that engineering was for me. I didn't know who to talk to. I didn't know what to read. I didn't know where to visit. I truly wish I would have had a resource to answer my questions about what an education, and later a career, in engineering was really like. This would have saved me time and stress in making my career decisions.

Over the decades, I've met many people in the same situation I was in who have many of the same questions, such as: high school students wondering whether they should enroll in engineering at university; engineering students pondering whether to get an industry position after graduation or do a master's or doctoral degree; master's and doctoral students in engineering dealing with concerns about applying for jobs in university, industry, or government; and engineers already in the workforce trying to figure out the daily ethical, financial, interpersonal, and technological challenges they face.

I've had the privilege of personally communicating with such folks over coffee, by phone, or by email. In retrospect, I wish I had a ready-to-go resource, like a book or audio or video or website, that I could also have passed on that addressed their concerns perhaps more effectively than I could. Sadly, many engineers are never really taught any career skills in university or even by their bosses in the workplace. They are forced to informally learn them as they go along, if at all. These career skills, however, are vitally important for a successful and satisfying career in engineering.

And so, this was my motivation in writing this modest little book. It's intended to address many practical questions about engineering regardless of specialty, such as university education, industry opportunities, business aspects, professional certifications, scientific fundamentals and methodologies, workplace communication and leadership, etc. As someone once said, I hope this book is as practical as a handle on a suitcase. Also, my goal is that the book gives readers a kind of grander poetic vision for who an engineer is and what an engineer does in helping design, build, and sustain human civilization.

This book is a series of personal conversation-style letters written to fictional people whom I call Nick and Natasha. It deals with their entire engineering career from early university education, then to the workplace, and then to retirement. It's not meant to be a scholarly textbook that addresses everything imaginable in a rigorous and systematic way. Nor does it presume or pretend to have all the right answers or tell people exactly what they should do in all situations. Rather, the letters are intended to be more like pearls of wisdom (I hope!) strung together that can help guide and mentor the next generation of engineers. With this in mind, let's now turn to the subject at hand.

Radovan Zdero, PhD, CEng, MIMechE

Part I

Basics

Letter 1 What's an Engineer?

Dear Nick and Natasha,

I'm pleased to start corresponding with you about the noble profession of engineering. Thank you for the privilege of letting me do so. This letter focuses on addressing a basic question: What's an engineer? This may seem like a trivial thing to do. But, I think it's best to start with the essentials before discussing more complicated ideas. Now, I know that many people in society have a rudimentary idea of what an engineer is. (Except for those people who still jokingly say we drive trains!) Yet, I think it's a very limited idea. And, unfortunately, I'm convinced that even many engineering students and working engineers have a rather narrow concept of engineering.

Popular media and culture often reinforce certain semi-accurate ideas about engineers. In the classic science fiction television series *Star Trek*, the chief engineer Scotty is an indispensable member of the crew, but he's often portrayed as a mechanic who spends all his time running around fixing things. One of the main characters in the popular comedy television show *The Big Bang Theory* is the engineer Howard Wolowitz who, although considered extremely competent at what he does, is frequently taunted for supposedly not being as smart as his physicist friends. And then there are all those engineers in books and movies who want to use technology to save, enslave, or destroy the world. Whatever merit these conceptions might have, I hope this letter helps somewhat to clarify and broaden your view of engineering.

A Basic Definition

Let's start with the word "engineer." It comes from the Latin word "ingenium," meaning "skill." It's related to the English words "ingenious" and "ingenuity." This refers to an ability to cleverly generate, understand, or explain a concept and/or provide a solution for a problem. But, how did this word come to describe our profession? The answer is, at least partially, historical.

As human civilization grew, certain people developed expertise in creating mechanical gadgets, building physical structures, and harnessing sources of power like the sun, wind, and water to solve practical problems. Different words were used to describe this class of people, such as artisans, craftsmen, master builders, mechanics, tradesmen, etc. Yet, the modern English word "engineer" probably dates back to 1325 to describe a person skilled at constructing military engines, that is, devices and machines.

Now, the *Oxford 3000 Dictionary* defines an engineer as "a person whose job involves designing and building engines, machines, roads, bridges, etc." In a strict sense, this is correct, but it's too narrow and doesn't even begin to touch on the wide variety of roles and responsibilities of an engineer. I've checked several other

DOI: 10.1201/9781003193081-1

dictionaries, and they all have narrow definitions similar to this one. Thus, I propose the following broader definition: *An engineer is a person who is trained to have a unique expertise for proposing and implementing innovative ways to solve practical problems using science and technology.*

A Jack of All Trades

As science and technology continue to advance into the future, the word "engineer" and the above definitions could become irrelevant. Even today, an engineer is much more well-rounded and multi-skilled than many realize. Such a person is often said to be a "jack of all trades." So, I'd like to present a series of aspects of what engineers actually do in today's world. Some individual engineering jobs embrace many of these aspects, while others embrace fewer. Nevertheless, some facets may become even more, or less, dominant in the profession in the distant future. With that said, here's my list of the top 14 aspects of engineering, in no particular order (see Figure 1.1).

An engineer is a *visionary*. They picture in their minds a desired future that is better than the present. That image drives them to think, speak, and act, so the desired future becomes a present reality.

An engineer is a *designer*. They conceive of small or large improvements on the form or function of older products (i.e. technology, technique, or process), as well as thinking of new ones.

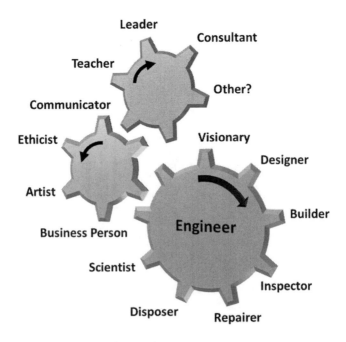

FIGURE 1.1 The many roles of an engineer.

An engineer is a *builder*. They use state-of-the-art machines, methodologies, and materials to fabricate or create initial prototypes and the final version of products.

An engineer is an *inspector*. They evaluate the quality and the performance of products, and then they make all necessary recommendations and changes.

An engineer is a *repairer*. They fix products that are not functioning properly according to the desired safety standards, technical specifications, user needs, and so on.

An engineer is a *disposer*. They store, discard, disassemble, and/or recycle products that are irreparable and/or out-of-date in a safe, legal, ethical, and effective way. This can also include disposing toxic materials and waste products.

An engineer is a *scientist*. They know how to use fundamental scientific principles and procedures to conceive, produce, and maintain effective products for their customers and society at large.

An engineer is an *artist*. They actively release their creativity to develop products that solve practical problems, while also appealing to human psychology and the physical senses.

An engineer is a *business person*. They make realistic decisions based on the "inflow" and "outflow" of finances for developing, marketing, selling, and distributing products.

An engineer is an *ethicist*. They consider their personal values, employer policies, professional standards, and society's laws and goals when developing effective solutions to practical problems.

An engineer is a *communicator*. They speak, write, and create resources to effectively communicate to other engineers, clients, customers, and broader society about their work as engineers.

An engineer is a *teacher*. They instruct, mentor, and train engineering students and working engineers, so they will have the knowledge and skill to bring effective solutions to practical problems.

An engineer is a *leader*. They formally supervise and/or informally influence small and/or large groups of people to accomplish goals in providing novel solutions to practical problems.

An engineer is a *consultant*. They offer expert advice to engineering and/or non-engineering clients, so they can make the best decisions about developing or purchasing products.

Strengths and Weaknesses

Every engineer is unique and potentially has the different aspects I just listed above in various combinations and measures, which can either help or hinder their career. Let's consider the legendary electro-mechanical engineer Nikola Tesla (1856–1943), as described by Margaret Cheney and Robert Uth in their book *Tesla: Master of Lightning*.

Tesla had hundreds of inventions and patents to his name. There's the brushless AC (alternating current) induction motor, fluorescent lights, radio, radio remote control, wireless power transmission, the Tesla coil, the "Egg of Columbus," just to name a few. He received many awards and was eventually honored by his peers, who named the strength of a magnetic field per square area by the scientific unit called the "Tesla" symbolized by the letter "T."

Yet, in my personal opinion, all Tesla's engineering accomplishments were due to his greatest ability, namely, having a clear vision of what he wanted to achieve. For instance, as a young boy he pictured a massive waterwheel being turned by the powerful waters of Niagara Falls to harness its energy. He told his uncle he would do it one day. And, true to his vision, he accomplished this feat 30 years later, when his revolutionary invention of AC power generation was implemented in 1896 by the monumental Niagara Falls Power Project. Later in the June 5th, 1915, issue of *Scientific American*, Tesla stated that "in this lies the secret of whatever success I have achieved. My imaginings were equivalent to realities."

But, like all engineers, Tesla had weaknesses. By all accounts, he was not driven by money in his work, but by an idealistic desire to improve the world. Ironically, this lack of business sense probably hurt his career. As a case in point, Tesla persuaded the wealthy financier J.P. Morgan to financially back his idea of building a massive tower that would transmit wireless communication signals across the entire world. By 1902, the 187-foot tower, power plant, and laboratory at Long Island, New York, were almost complete. But, when Tesla asked Morgan for extra funds and revealed the true nature of the project—to provide cheap power to the entire globe—the profit-driven financier abruptly backed out of the plan. With dwindling funds, Tesla had to shut down the project in 1905. The tower was finally demolished in 1917.

So, What's the "Take Home" Message of This Letter?

The gist of this dispatch is that an engineer not only designs, builds, inspects, repairs, and disposes of devices, machines, processes, etc., but also potentially has an array of other roles and responsibilities that are incorporated into the job. Of course, a day in the life of a particular engineer may be very different than their colleague's. Some will have more of these facets to their job, while others will have fewer facets. But, on the whole, the engineering profession includes a broad sweep of duties and experiences, making it much more holistic than many people think. Now, whether you're just thinking about studying engineering at university or a technical school, you're currently a student engineer, or you're a working engineer, there are many opportunities for professional growth. I hope this letter encourages you to explore new facets of what it means to be an engineer.

Sincerely,

R.Z.

Letter 2 The Different Types of Engineers

Dear Natasha and Nick,

I trust this letter finds you doing well. This one took a little longer than I expected because it required a bit more research on my part to make sure it was accurate. In any case, you wanted to know about the different kinds of engineers that exist, so you could have a bigger picture of the profession as a whole. And so, I've tried to do my best in providing the following response.

There are, what I like to call, the "Classic 4" engineers. Regardless of which university it is, whether large or small, famous or unknown, the vast majority of them will offer, at bare minimum, chemical, civil, electrical, and mechanical engineering programs. One main reason for this is historical. Basically, the technological needs of human civilization after the Agricultural Revolution 10,000 years ago were mainly met by civil engineers (who built roads, waterways, and palaces) and mechanical engineers (who built farm tools, household devices, and weapons).

But, because of the Scientific Revolution of the 1500s and 1600s, as well as the Industrial Revolution of the 1700s and 1800s, there was a growing desire to understand and harness the power of phenomena like electricity (hence, electrical engineers) and atoms, molecules, liquids, and gases (hence, chemical engineers). And then other engineering specialties that came along over time were usually spin-offs from these "Classic 4." Let's first take a look at these 4 in some detail (see Figure 2.1).

Chemical Engineer

To become a chemical engineer, the typical core courses you would take in university include fluid mechanics; fluid dynamics; mass, heat, and energy transfer; mixer design and engineering; process control and optimization; reactor design and engineering; statistics and probability; and thermodynamics. You will also usually have the opportunity to register in several technical and perhaps even non-technical arts, humanities, or social science courses of your choice. A few of these could be from the other "Classic 4" or spin-off disciplines, so you will be a more holistic chemical engineer. Once you graduate and get into the so-called real world, as a chemical engineer your work could be in one of several major industries, such as composite materials, food and drink, metal processing, microbiology, nanotechnology, nuclear energy, oil processing, pharmaceuticals, polymer synthesis, etc., just to name a few. So, for instance, chemical engineers optimize and manage complex systems and processes that deliver products to society. This includes oil refineries that have a vast array of components arranged almost like an assembly line that processes crude oil so it can be transformed into fuel for vehicles and

DOI: 10.1201/9781003193081-2

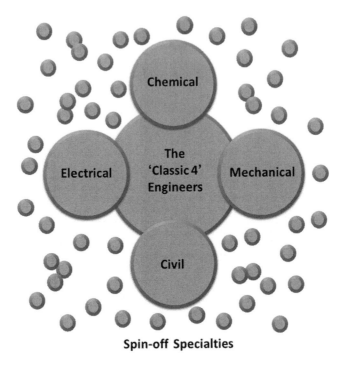

Spin-off Specialties

FIGURE 2.1   The "Classic 4" engineers and their spin-offs.

machines. Similarly, food processing plants are more complicated than many people think and, thus, require optimization of the assembly line that takes raw potatoes from the field and transforms them into ready-to-enjoy potato chips.

Civil Engineer

To become a civil engineer, the typical core courses you might study in university are concrete materials; environmental science; fluid mechanics; geology; highway materials; hydrology and water resources; mechanics; pavement materials; soil properties; solid waste management; strength of materials; structural design; surveying; traffic operations and management; transportation planning; and wastewater management. You will also often have a chance to study a few technical and maybe even non-technical arts, humanities, or social science courses of your choosing. Some of these may be from the other "Classic 4" or spin-off disciplines, which are intended to help you be a more holistic civil engineer. Once you complete your studies and find yourself in the real world, as a civil engineer your projects could involve designing, building, inspecting, and/or repairing bridges, buildings, city layout plans, dams, railways, roadways, soil erosion, surveying, suspension cables, traffic flow, water works, wind mills, and so on. It's emotionally moving to consider just a few of the many iconic structures—old and new—around the globe that typify all the best of what civil engineering has to offer. From the pyramids of the ancient

Egyptians to the aqueducts of the ancient Romans. From the Cathedral of Notre Dame in France to the underwater and underground Chunnel that connects Britain with continental Europe. From the sail-like roofs of Australia's Sydney Opera House to the needle-like profile of Canada's CN Tower.

Electrical Engineer

To become an electrical engineer, the typical core courses you'll probably take in university include circuits and circuit design; computer architecture or hardware; computer coding or programming; electromagnetics; signal processing; statistics and probability; and power generation and distribution. You'll probably have the option to also enroll in several technical and non-technical arts, humanities, or social science courses of your choice. Several of these may be from the other "Classic 4" or spin-off disciplines, to assist you in becoming a more holistic electrical engineer. Once you finish university and join the workforce, as an electrical engineer your work could involve several areas that can be classified as low voltage (e.g., biomedical implants, computers, smartphones, etc.), mid voltage (e.g., aircraft, appliances, robots, satellites, solar panels, vehicles, etc.), or high voltage (e.g., hydro or nuclear power generation and distribution stations). Whether we are at home, at work, or out in public somewhere, we are continually surrounded by the marvels of electrical engineering. For instance, mighty power stations around the world, like at Niagara Falls, harness the force of cascading water and transform it into electrical power that energizes our cities, workplaces, and homes. Similarly, countless medical devices in hospitals and clinics, like pacemakers and ultrasound machines, are vital tools used by healthcare professionals to effectively diagnose and treat the illnesses and injuries that so easily felled our ancestors. And high-tech space telescopes search the cosmic ocean for answers about our solar system, our galaxy, and our universe.

Mechanical Engineer

To become a mechanical engineer, the typical core courses you might take in university are statics (i.e., forces and displacements of stationary objects and structures); dynamics (i.e., forces, displacements, velocities, and accelerations of moving objects and mechanisms); fluid mechanics; fluid turbulence; heat transfer; machine design; mechanical design; mechanical vibrations; strength of materials; and thermodynamics. You will likely be able to enroll in several technical and even some non-technical arts, humanities, or social science courses of your choosing. A couple of these may be from the other "Classic 4" or spin-off disciplines, in order to make you a more holistic mechanical engineer. Once you finish university and find yourself in the real world, as a mechanical engineer your work could involve designing, building, inspecting, repairing, and/or disposing of machines, medical implants and prosthetics, nuclear reactors, pumps, robots, sports equipment, turbines, vehicles, weapons, just to name a few. The beauty and benefits of mechanical engineering are clearly displayed—among the countless possible examples—in the 16.8-m long robotic Canadarm that's attached to the International Space Station orbiting 400 km above the Earth, the high-speed trains that use magnetic levitation technology to transport people and goods at about half the speed of sound, and the massive number of shoulder, hip, and knee implants inserted into patients every year around the world to help them get back to living their lives actively without pain.

Engineering Spin-Offs

Now, as mentioned briefly above, a huge array of engineering spin-offs of other specializations eventually emerged over time from the "Classic 4" to keep pace with scientific advances and societal needs (see Table 2.1). Sometimes one or more of the "Classic 4" disciplines partnered together deliberately to launch a spin-off discipline, but sometimes this process was unplanned and accidental. Thus, some of these spin-offs are taught at the university as a specialty within one or more of the "Classic 4" engineering departments.

Yet, many spin-offs have become unique in their own right as separate and independent engineering programs; this will vary greatly from university to university. Quite often, however, the spin-off might not be even be recognized formally as a specialty within an engineering program at university, but it's only called by its particular name within a specific sector of industry. You can probably guess what each spin-off is about just from its name. Otherwise, we'd be here all day going over them one by one in detail. Keep in mind that the spin-offs I mention aren't all the ones that exist, but they are some of the most popular and well-known. As science and technology continue to advance, other specialties will emerge that no one's even thought of yet.

Career Dead End?

But, you may ask, doesn't such specialization lead you to a career dead end? I personally don't think it does. You're not necessarily forever stuck in doing only

---

TABLE 2.1
Engineering Spin-offs from the "Classic 4"

Aerospace	Geological	Nuclear
Agricultural	Geomatics	Petroleum
Automotive	Industrial	Railway
Biochemical	Manufacturing	Robotics
Biological	Marine	Software
Biomedical	Materials	Space
Computer	Mechatronics	Systems
Electronics	Metallurgical	Water
Environmental	Military	Wind
Forest	Mining	Other?

---

one kind of task for the rest of your career. The reason is that many engineering disciplines overlap because of the unplanned way these specialties emerged over time. So, if you're already one type of engineer, you may be able to switch to another specialty because you already know something about it. In these cases, it's not too difficult to transition your career from one kind of engineering to another. People actually have done this due to changes in their circumstances, opportunities, and interests.

For instance, I did my master's degree in mechanical engineering with a research thesis focused on the aerodynamics of wind flow around overhead electrical power lines that accumulate ice on their surface during winter. But, then I did my doctoral degree in mechanical engineering with a research thesis on using medical ultrasound technology to analyze the performance of total knee replacements under clinical-type stresses. In my case, my home department was mechanical engineering in both instances, thus, it probably made it easier for me to change my research focus.

However, I know a colleague who did a master's degree in materials engineering, but then became a university professor in mechanical engineering. Other colleagues I know did their master's degrees in aerospace or mechanical engineering, and then found work as biomedical engineers in industry. There are others I know who were educated and trained as mechanical engineers, but who became university professors in systems engineering departments. These real-life situations clearly show that you can even successfully change engineering disciplines (within reason, of course) because of the many transferable skills and knowledge that all engineers share in common with one another.

So, What's the "Take Home" Message of This Letter?

I hope my letter has been helpful in answering some of your specific questions about the different types of engineering that exist. And perhaps, it has given you some ideas about what kind of engineer you want to be or what kind of career changes you might consider in the future. I also hope that you can now appreciate how the engineering profession as a whole impacts almost every aspect of people's daily lives and society at large. And that you too can be part of this great profession.

Best regards,

R.Z.

Letter 3 A Brief History of How Engineers Built Civilization

Dear Nick and Natasha,

I hope everything is going according to plan for you personally and professionally. Lately, I've been thinking about how easily we engineers get bogged down with our studies, teaching, or research at the university or with our work in industry or the government. And rightly so, since we do have a responsibility to ourselves and society to get the details right in our designing, building, inspecting, repairing, and disposing. But, I think we sometimes miss the big picture of what we are doing and why it's important. Although not as celebrated as political figures, military leaders, or entertainers, it's obvious to anyone that engineers have always played a critical role in society.

Just take a look at history and our modern lives. Almost every aspect of society is impacted in some way by the marvels of technology. So, although I don't pretend to be a professional historian, I'll do my best to tell you the story of how engineers built civilization. And then I'll try to highlight some key themes that I personally think have always been vital to the onward march of technological progress. If you want to find out more, I recommend the 1960 classic book by L. Sprague de Camp called *The Ancient Engineers*, which I consulted for much of the information below from ancient times to the Renaissance. Now, our story begins thousands of years ago after the invention of farming (see Figure 3.1).

The Agricultural Revolution (c.8000 BC)

The invention of farming and the first domestication of livestock happened in Mesopotamia (roughly today's Iraq) and was critical to the rise of engineering. It allowed previously nomadic people (i.e., hunters and gatherers) to settle down in an area, grow surplus crops and animal products, and sell their goods to others. This allowed these other folks to shift away from farming and raising livestock themselves, which permitted them to organize society into population centers like villages and cities. Some of these village and city dwellers became craftsmen and tradesmen as they designed and made new implements for household, farming, public, and military use. Thus, technological innovation was born, and these craftsmen and tradesmen became the first engineers.

DOI: 10.1201/9781003193081-3

FIGURE 3.1 Engineering throughout history.

The Ancient Mesopotamians (c.3000 BC–c.500 BC)

The Mesopotamians of the distant past lived in a large flat hot desert-like plain. They built houses that were laid out in a square pattern with an inner court where the owner sat in the shade; there were no "outside windows," thus, blocking the sun's heat. This house design also gave some protection and privacy against burglars and meddlesome government agents. Also, water canals were built for crop irrigation, and levees controlled river flow. Some villages grew into larger towns and cities with unplanned haphazard layouts creating problems of congestion, sewage, transport, and law enforcement. Ziggurats were step-like pyramid temples built for religious purposes, like the most famous one dedicated to the god Marduk, which may be the Tower of Babel mentioned in the Bible.

The Ancient Egyptians (c.3000 BC–c.300 BC)

The Egyptians of antiquity made technological advances often through the leadership of their Pharaohs, since they had time, resources, wealth, and power. They built public works, military equipment, monuments, temples, and pyramids using clay, mud, straw, and stone. To build the pyramids, new technologies had to be invented, like water-swelled wooden wedges that made fracture lines to quarry huge stones, large barges for transporting stones down the Nile river, and large numbers of men conscripted into forced labor to haul stone blocks on milk-lubricated wooden sled-like devices across the ground. Imhotep, the first engineer and architect known in history by name, was one of those skilled men employed by his Pharaoh to build a pyramid. Although much of this was intended for the dubious purpose of boosting the egos of the Pharaohs, the new knowledge was useful to later generations.

The Ancient Greeks (c.700 BC–c.30 BC)

The Greeks of olden times made huge contributions to science and technology, but I'll just list a few here. Blacksmiths made sturdier soldiers' armor that was covered and reinforced with bronze, iron, or steel. Kings employed artists and architects to incorporate the first use of metal structural members in buildings, shrines, and statues. Archytas of Taras, who was a friend of the famous philosopher Plato, invented the screw. Aristotle advanced science and technology by founding the scientific method that combined theorizing with gathering experimental data. Also, Aristotle and/or Straton of Lampsakos wrote *Mechanika* (i.e., *Mechanics*), the world's first engineering textbook on gears, levers, pulleys, rollers, slings, wedges, and beam balance for weight measurement.

Archimedes of Syracuse is considered the greatest engineer of the ancient world. He made major contributions to mathematics (e.g., calculating Pi more accurately), general engineering (e.g., principles of the lever, pulley, screw, and wedge), and military engineering (e.g., cranes that dropped heavy stones onto attackers scaling the city wall). King Alexander the Great commissioned the architect and engineer Deinokrates to build the city of Alexandria, which was home to the great Library of Alexandria (which had up to 750,000 scrolls) and the Museum of Alexandria (which had laboratories and lecture halls and was the base for many scholars, so it was essentially the first university in the ancient world). Ktesibios of Alexandria invented metal springs made of iron or bronze, musical pipe organs based on hydraulic and pneumatic action, water clocks for telling time, and water pumps for raising water to a height.

The Near and Far Eastern Civilizations (c.500 BC–c.1100 AD)

In India, the rise of Buddhism after 500 BC saw the construction of many temples and monasteries. The emergence of a Hindu empire in 300 AD resulted in cave temples carved out of mountainsides and large detached stone temples. And there were craftsmen skilled at ironwork for making chain bridges, dowels, ingots, and pillars.

In China, King Tsin ruthlessly united all the warring Chinese states for the first time in history into one empire in about 220 BC. He then embarked on an ambitious construction campaign that included many 60-ton statues of himself dotting the empire and the 2250-mile long Great Wall of China to keep raiding bands of Mongols at bay. Over the centuries, the Chinese also made contributions to communication (e.g., book printing), time keeping (e.g., clockworks), navigation (e.g., magnetic needles and/or compasses), and war (e.g., gunpowder and primitive rockets).

In the united Arabic empire and its later successor states (632 AD – c.1100 AD), Christian and Jewish scholars were sometimes employed to translate ancient scientific texts into Arabic. Then, the Arab scholars took these ideas and made further contributions to science (e.g., astronomy, chemistry, mathematics, medicine), which they utilized to develop new technology. The Arabs also made innovations in architecture (e.g., mosques and minarets) and the military fortification of city walls (e.g., trap door structures that could be opened to shoot arrows, drop rocks, and pour boiling oil onto the heads of the enemy). They also built canals, dykes, gears, moats, roads, and water clocks.

The Ancient Romans (c.400 BC–476 AD)

During the rise and fall of their vast empire, the Romans mainly designed and built things that were of practical use to civil society. This includes things like apartment buildings, arenas, aqueducts, bridges, circuses, forums, fountains, harbors, military equipment, naval headquarters, public bath houses, roads, sewers, temples, theaters, town halls, and water storage tanks. These construction projects were often spurred on by times of peace, the availability of cheap labor, and the hunger for personal glory by Roman emperors. But, they gave little attention to pure science research.

Also, there was not as much attention paid to creating new mechanical devices for household or public use, since this did not bring much public recognition to a Roman emperor's achievements. Ironically, the layout of the city of Rome itself, like most ancient cities, grew in a rather unplanned organic way, resulting in a hodgepodge of various types of buildings next to each other, winding alleys and roads, and congestion of street traffic. When it came to private homes, upper class people were able to obtain central heating, glass windows, and even piping to bring in water from water storage tanks that were kept filled by a network of aqueducts.

Several important technical writings appeared at this time. Marcus Vitruvius was a Roman architect and artillery engineer who wrote a 10-volume work called *De Architectura* about acoustics, architecture, building materials (brick, concrete, masonry), construction methods, interior decoration (paint, plaster, mosaics), measurement methods, military equipment, town planning, and water supply. And Heron of Alexandria (then part of the Roman empire) wrote several important treatises on measurements, mechanics, military weapons, optics, pneumatics, surveying, water clocks, etc.

The Medieval Byzantines (476 AD–1453 AD)

The Byzantines inherited the eastern half of the old Roman empire, whose western half collapsed in the year 476 AD because of barbarian invasions. The relative peace, prosperity, and political stability of Byzantium allowed science and technology to advance.

The Byzantine emperor Justinian (ruled 525–567 AD) was a great builder who met the needs of the faithful by constructing churches, which also functioned like town halls where judges heard court cases, workers guilds could meet, and city officials would conduct business. He also built fortifications like towers and walls for the cities of the empire, roads through mountainous areas, aqueducts to carry water into the capital city, and overflow channels to protect the capital city from flooding.

But, by far Justinian's greatest engineering achievement was the construction of the Church of Saint Sophia in the capital city of Constantinople, whose central dome rose 160 feet high and was so lavishly decorated inside with mosaics and colored marble that it was instrumental in converting King Vladimir of Russia to Orthodox Christianity in the year 987 AD.

Some churchmen promoted science and technology, but most clergy preferred theological matters. Craftsmen made mechanical devices, yet did not develop them further into practical machines of any significance. One important writing on military technology was by Heron of Byzantium (also known as Heron the Mechanic) titled *Book of Machines of War*. It provided technical details on the bore, catapult, flying bridge, mantlet, pneumatic scaling ladder, ram, and sambuca.

The Western Europeans (c.900 AD–c.1600 AD)

The newly-independent people and nations of western Europe inherited the western half of the old Roman empire, which collapsed in the year 476 AD. They finally emerged from the political chaos, mass migrations, and scientific and technological ignorance of the so-called Dark Ages for several reasons.

They imported Greek scholarly manuscripts from the eastern half of the Roman empire that was still intact (i.e., Byzantium) and scholarly manuscripts from the Arabic empire. These documents were then translated into Latin by Christian monks and Jewish scholars.

Scholars wrote works on science and technology that brought together previous knowledge or promoted original ideas, such as the many writings of the Christian monk Roger Bacon (1214–1292 AD). Not surprisingly, new or improved inventions emerged, like arches and trusses for architecture, canals, cast iron, castles, cathedrals, eyeglasses, oil lamps, paved roads, plows, personal armor for soldiers, printing presses, sewers, ship building, stone bridges, water mills, wind mills, etc.

The Renaissance (1400s and 1500s AD)—which means rebirth or revival—saw the rise of a number of key events. There was a renewed interest in classical Greek and Roman knowledge. The first formal patent law was established in 1474 AD in Venice. The first research institutes and industrial expositions emerged. And there was an upsurge of professional architects, engineers, and scientists. One of these important figures was Leonardo Da Vinci (1452–1519 AD). His notebooks were filled with original drawings and blueprints for axles, gears, levers, pulleys, ratchets, and springs, as well as bridges, canal diggers, city layout plans, fortresses, hoists, human anatomy, human-powered flying machines, mirror grinders, needle grinders, odometers, screw jacks, siphons, water wheels, etc.

The Knowledge "Revolutions" (16th–19th centuries)

A series of intellectual upheavals occurred from the 16th to 19th centuries AD that greatly accelerated the march of science and technology to eventually create our modern world. The Scientific Revolution (1543–1687) really got going with the publication of Nicolaus Copernicus's *On the Revolutions of the Celestial Spheres*, which proposed that the planets revolved around the Sun, not the Earth. This important period came to a close with the publication of Sir Isaac Newton's *Principia*, which introduced and mathematically explained the laws of gravitation and motion. And it saw the establishment in 1662 of the world's oldest surviving scientific organization, the Royal Society.

The Industrial Revolution (1769–1829) was launched in 1769 by James Watt who perfected the steam engine, which replaced muscle-power with steam-power. This radically transformed many industries in Britain, like cotton spinning, cloth weaving, flour milling, and paper milling, as well as improving marine transportation. By the end of this critical period in 1829, the British engineer George Stephenson designed the first practical steam-powered train called "The Rocket." The railway constructed the following year to connect the cities of Liverpool and Manchester signaled the start of rapid mass transit. In the aftermath, some of the world's oldest formal engineering societies that still exist were created in Britain, such as civil (1828), mechanical (1847), and electrical (1871).

The Technological "Ages" (20th–21st centuries)

In more recent times, a quick succession of technological breakthroughs were spurred on by the doubling of human knowledge every 18 months or so. The *nuclear age* saw nuclear power plants that supplied massive amounts of cheap energy to cities and industries, but it also brought with it the specter of nuclear weapons, war, and annihilation. The *space age* saw humans orbiting the Earth, landing on the moon, living on the International Space Station, and planning to colonize Mars, yet it also led to powerful nations racing to dominate and weaponize space. The *computer age* saw the almost ubiquitous use of helpful new technologies like laptops, tablets, and smartphones, but it was also fraught with concerns about Artificial Intelligence machines and robots eventually becoming sentient and then turning on their human creators in a Frankenstein's monster scenario. The *internet age* saw the rise of global information sharing, communication, and connectedness on an unprecedented scale, yet it also allowed malevolent actors to spread their dubious and dangerous content across the planet through the World Wide Web. And that's just a tiny sample of the rapid pace of change from the past century.

What Lessons Can We Learn From Engineering History?

I'd like to suggest several lessons from the past that I think are particularly important to apply today, but there may be different ones you could also tease out that I may have missed.

First off, I'd emphasize that engineering is founded on *science*. By this I mean that the laws of physics, chemistry, and biology that regulate the universe dictate what engineers can achieve practically. So, let's make sure we are extremely familiar with, and even take part in discovering, the basic scientific principles that govern the devices, structures, techniques, or processes with which we work.

Next, engineering is fueled by *practical necessity*. By this I mean that the daily needs and occasional crises of society are opportunities for engineers to come up with innovative solutions. So, let's always be mindful of all the many possibilities that exist for us engineers to use our skills to improve society.

Also, engineering is not revolutionary, but *evolutionary*. By this I mean that most, but not all, new technological developments are small improvements on existing older technology, yet these can still grow over time to make a big impact on society. So, let's not minimize the work we do, no matter how seemingly ordinary or unimportant it may seem to us at the moment.

Let me now suggest, whether we like it or not, that engineering wheels are greased by *money*. By this I mean that the successful development of new technology always requires a source of funding, whether from the top or the grassroots. So, let's always be aware of financial needs and opportunities that can bring our innovative ideas into reality.

Finally, I think engineers must be true to their personal and professional *values*. By this I mean that we engineers need to be careful in deciding what projects we are and are not willing to accept because of the potential negative consequences to society; just because we can do something, doesn't mean we should. So, let's be certain to make use of our own personal ethics and professional standards as guidelines in pursuing our work.

So, What's the "Take Home" Message of This Letter?

I hope this note gives you a strong sense of being part of a long unbroken line of engineers that stretches deep into the misty past as civilization was just getting started, but that also reaches far into the bright future of what civilization could be. Even so, I can't help asking some questions: What does the future hold for engineers and engineering itself? What will be the new fundamental scientific discoveries in biology, chemistry, mathematics, and physics that engineers can then turn into practical applications? What kinds of novel materials, fabrication techniques, and energy sources will we need to develop? Will unforeseen societal problems, natural catastrophes, and manmade disasters compel engineers to quickly find solutions before it is too late? What new uncharted frontiers will beckon to us for communication, exploration, and travel, but which will require advanced technologies that we haven't even thought of yet? And, perhaps most importantly, will we have the moral courage to engage these challenges and opportunities while staying true to our values? Alas, only time will tell.

All the best,

R.Z.

Part II

Preparation

Letter 4 To BE or Not to BE: Studying Engineering in University

Dear Natasha and Nick,

I'm glad about your continued interest in becoming engineers. That's a noble goal. But, first you have to get a BE (bachelor of engineering) degree from a university or technical school. Equivalent degree designations may be BEng (bachelor of engineering), BEngSc (bachelor of engineering science), BASc (bachelor of applied science), or something similar. Once you get a BE, you can legally put those letters after your name on your office door, business cards, documents, email signatures, letters, etc.

More importantly, enrolling in an accredited university engineering program will give you the systematic and comprehensive intellectual preparation needed. As with a house or other building, the foundation is the most important part that the rest of the structure will rest upon. If your foundation is faulty, the structure will eventually start to show cracks, spring leaks, and experience other flaws that may eventually cause it to totally collapse.

Others may disagree, but I believe a high quality engineering program should have a vision statement that goes something like this: "To equip the next generation of engineering students with an intellectual foundation, practical know-how, and problem-solving skills, so they can produce effective solutions for real-world problems." Having said that, you should know something about the experience of a typical university engineering education. That's what this letter's all about.

Because I can't possibly deal with every conceivable question or situation in this letter, I'll focus on 5 key tasks required in a typical BE program and 5 key tactics for success so you'll do well in your studies (see Figure 4.1). By the way, I've numbered the tasks and tactics in no particular order, so don't try to match the tasks with the tactics. Also, some tactics are applicable to several tasks.

Some General Thoughts on BE Degrees

Let me start off by making some general observations about BE degrees before we get into the specific tasks and tactics that I'd like to focus on.

First off, you may wonder if it matters *which university or technical school* you attend to get your BE. If you're fortunate enough to go to a famous university, then this may give potential employers an extra positive impression of your intelligence

DOI: 10.1201/9781003193081-4

FIGURE 4.1 The tasks of a BE program and the tactics for success.

and abilities. So, they may take a closer look at your resumé when you apply for a job. But, other than that rare situation, in my opinion, it's not going to matter very much to employers what university or technical school you attend. Other things will matter more. The grades you achieved in your courses will matter, since this indicates your general intelligence. The specific courses you studied and the skills you acquired will matter, since they show your suitability for the particular job. Any prior relevant volunteer or job experience you have will matter, since this will make it easier for employers to train you. And your personality and performance at a job interview will matter, since employers want to know if you are easy to get along with and can work well with others on a team.

Some universities offer the first BE year as *general engineering*, whereby all students take the same courses (e.g., chemistry, design methods, mathematics, mechanics, physics, etc.) to give them the same background preparation. The benefit is that it gives students some time and experience to better decide which engineering specialty (e.g., biomedical, chemical, civil, electrical, mechanical, nuclear, software, wind, etc.) they want to pursue in the upper years. Other universities, however, expect students to immediately enroll in an engineering specialty in the first year. The benefit is that students will have an additional year of specialized courses in their discipline upon graduation. You have to decide if one or the other approach is the best when you decide which universities to apply to.

Keep in mind that every year of a BE program may have a *different difficulty level*. For instance, the first year of my BE was expected to filter out weaker students, so they would quit engineering. I know this because at a public student orientation event, an engineering professor told us students that he expected that one-third of us would quit engineering before the first year was over; many, in fact, did quit. This was not surprising since the first year was the most difficult of all 4 years of the program—it had the most number of courses in the curriculum, followed by the third, the second, and the fourth year, which was by far the easiest. You should know that most university engineering programs take 4 years to finish. However, there may be exceptions in certain schools that require completion of a work practicum or that offer an additional business management component which can extend the program's duration.

Moreover, typical BE programs have several different *types of courses*. The most important and numerous ones are mandatory and are sometimes called "core" technical courses. They are important because they will teach you all the fundamental knowledge and skills needed for your particular engineering specialty, whether that specialty is biomedical, civil, mechanical, software, etc. But, you'll also likely be able to take a few "elective" technical courses that are from another engineering specialty in order to make you a more well-rounded engineer. They are important because you may end up working with different types of engineers at different times in your career, so it's good to have a broader understanding of engineering. On top of that, you'll probably have the opportunity to take a few "elective" non-technical courses from the arts, humanities, or social sciences in order to make you a more well-rounded person. They are important because they can give you insights into human psychology and behavior, so that you can design products that are more optimally suited to human wants and needs.

Although a BE program will keep you busy studying, consider *volunteering* for an extracurricular activity. There are plenty of volunteer opportunities available for engineering students. These could involve non-engineering groups, such as charitable organizations in the community, formal or informal sports teams, student clubs on campus, etc. Or, if you want to do something related to engineering, you can join the local student chapter of a professional engineering society, you could become a university student ambassador who gives speeches about engineering to high school students, you might want to be on the organizing committee of a science and technology fair, and so on. These activities can help you relax, take your mind off the stress of your BE studies, and help you live a more balanced life. It also gives you the chance to meet new people in order to start building your professional network of contacts, which can potentially open up all sorts of career opportunities for you in the future.

It's also encouraging to know that you can begin work as a *junior engineer* immediately after getting a BE degree. But, you still need to write a certification exam, do an interview, and/or get a year or more of hands-on experience before you become a legally recognized "professional engineer" who can lead projects that potentially affect public health and safety. Many countries recognize the legitimacy of BE degrees obtained in other countries, so you may have some options to work abroad. Be aware that, in some countries, a technician and technologist can train at a technical school to get a certificate and then eventually do work that sometimes overlaps with an engineer. However, that diploma will still not carry the recognition, rights, responsibilities, or rewards of a BE university degree.

Now, you may be wondering if you'll really need to *know all the* "*stuff*" you'll learn during a BE degree—like concepts, formulas, procedures, techniques, and tools—after you get a job. The answer is, probably not. But, you'll need to know some of them, depending on the job you get. Let me explain. A BE is meant to give you depth and breadth in skill and knowledge in a particular engineering specialty (e.g., aerospace, civil, mechanical, software, etc.). This is meant to prepare you for all the different potential jobs you might get within that engineering specialty. Yet, every engineering specialty can be further divided into subspecialties that deal with particular tasks. So, there's no way the university, the professors, or even you can predict exactly what job you will get and what skills and knowledge you'll need for the day-to-day tasks of that job. For your own sake, it's good to absorb as much as you can, since you'll probably use different parts of your BE education at different times for different jobs requiring different tasks.

And finally, a BE degree is *highly prized* no matter where you go in our technologically-driven world. So, don't underestimate the positive impression that your degree can make even on non-technical people. There is a stereotype about engineers that you should leverage to your own advantage. Clients, colleagues, customers, employers, friends, and strangers may often automatically assume that you are intelligent, hardworking, responsible, detail oriented, able to deal with numbers, able to fix computer problems, and able to fix mechanical devices. The transferable nature of your knowledge, experience, and skills can open up all sorts of doors of opportunity for you. These opportunities could include things like switching to another engineering specialty, becoming a board member of an engineering firm or other type of company or organization, pursuing a career in another field, and so on.

Task 1. Lectures

Traditionally, BE course lectures have always been presented in-person live by a professor to students gathered in a lecture room at a particular time for a series of one-time lecture events. There are benefits to this approach. It can help build a good professional rapport of professor-to-student and student-to-student. The professor can make quick on-the-spot adjustments to the lecture material depending on how students seem to be responding. And students can ask questions at any point in the lecture to get an immediate response from the professor. But, there are drawbacks to this approach too. Everybody needs to gather at the same time in the same physical location. It does not allow students to watch the lecture multiple times or at their own convenience. And it makes little allowance for professors or students who get sick or who have scheduling conflicts.

With the advent of new audiovisual and internet technologies, several alternative options are available. Course lectures can be given live online over the internet to very closely simulate the traditional in-person experience along with all its benefits, but without the limitation of everyone needing to meet together in the same physical location. Course lectures can also be pre-recorded and then posted online. This lets students watch the lectures any time from any place as many times as they like and without any distracting background noise due to conversations and latecomers. Another philosophical shift is to incorporate a flip-the-classroom segment into the traditional or online lecture. This method reserves a portion of the lecture time for dividing students into small groups, so they can work on engineering design problems, perform calculations for practice problems, and so forth.

Task 2. Assignments

A BE course will require you to do homework that you will usually complete on your own free time when not attending or watching a lecture. Then, you will submit them to the professor or teaching assistant. They are almost always graded and will count towards your final course grade. How many assignments you will be asked to complete and how much each of them is worth can vary a great deal. Sometimes you will be asked to complete these on your own without consulting other fellow students, but other times the professor may allow it. Think of assignments as mini projects that are really meant to help you learn as you go along through the course, rather than just numerically evaluating your performance. Assignments could involve problem solving by performing calculations, inspecting and reporting on a failed gadget that you are given, providing small essays or verbal responses to qualitative questions, writing or verbally giving a critical review of a published research article, or any number of similar tasks depending on the course and the creativity of the professor. Keep in mind these could be completed in-person, on paper, via email, or using online technology over the internet.

Task 3. Quizzes, Tests, and Exams

A key part of a typical BE course is the quiz, test, or exam (QTE), which is a quantitative evaluation of your individual understanding of the course material. A quiz is often only worth a few percentage points of your final course grade, whereas tests are worth more and exams are worth the most. You will complete a QTE on your own without consulting other students. You will complete a QTE in-person in a room or lecture hall, or virtually online, but will somehow be monitored to make sure no one is cheating. You will often not be allowed to access any outside resources like a textbook or the internet while completing a QTE, but there are sometimes exceptions to this. You will solve some problems by performing computations, writing short answers, giving live video responses online over the internet, and/or providing in-person responses to the examiner. Now, keep in mind that although the professor may occasionally introduce some small twist to a question or problem in a QTE to see which students can extend what they've learned to a slightly new situation, the typical professor is not going to try to deliberately fail you or trick you by introducing a brand new major concept. But, if you do have a professor that routinely adds brand new concepts to a QTE, then it would be a legitimate reason for students to formally complain to the professor and even to the administration in the engineering department.

Task 4. Hands-on Labs and Projects

Engineers not only have to work with their minds, but also with their hands. To this end, BE university programs typically offer practical lab exercises to give you hands-on training for how to design, build, inspect, repair, and even dispose of devices, equipment, machines, structures, or processes. You should know several things about labs.

You will sometimes work with a small team of other students, so you need to hone your social interaction skills. Sometimes an individual course will have a lab exercise as part of the syllabus, whereas some engineering programs combine all the lab exercises into a separate independent course. The labs will be run by the professor, a teaching assistant who is doing their master's or doctoral degree in engineering, or by a lab coordinator who is employed full-time by the university. A typical lab exercise may only take 1 to 3 hours to complete. You may notice that the equipment used for the labs may be older technology, since it would get rather financially expensive to keep updating the equipment each term or year for each new group of students. You will eventually be required to submit to the professor a final written lab report either individually or as a group. Labs and projects may need to be completed in-person in a physical location, or perhaps using some sort of online technology over the internet.

Although major course projects have some of the above things in common with labs, they are different in several ways: (i) they are not limited to a couple of experimental apparatuses located in a dedicated room on campus; (ii) there is much more time and work involved in a project, perhaps over a period of several months; (iii) you will need to work much more independently, although you certainly should ask the professor for help with any problems; and (iv) you will submit a final pre-recorded or live video presentation or a live in-person presentation to an audience composed of the professor, fellow students, and maybe even industry folks who may be the judges grading your presentation.

Task 5. Work Practicums

Some, but by no means all, BE programs at university offer or require students to work in industry for a time in order to obtain their degree. The total length of time and how it is distributed throughout the duration of the engineering program will vary from university to university and from country to country. In some cases, this will be built into the engineering program so that it still lasts a total of, say, 4 years, but in other cases this will extend the program and delay your graduation.

But, the intention, obviously, is to give students hands-on experience as an engineer in the "real world." After all, there is only so much you can learn from textbooks and even hands-on labs or projects in a controlled university setting. Even so, many engineering firms offer on-the-job training for the specific task you will be involved in as an employee and, thus, doing a work practicum during your BE studies may not necessarily be of any great benefit to you in the long-term.

For instance, often the company with whom you did your work practicum will hire you as an engineer because they know you and your abilities. Also, in this situation, you are already trained and, thus, you will very quickly start contributing to the company's goals and will make your boss happy because of it. But, even if that company does not hire you, the hands-on experience you gain will look very impressive on your resumé when you start applying for other engineering positions. Moreover, the people you meet during your work practicum will be a good start to building your own network of personal and professional contacts that can open up all sorts of opportunities for you sometime in the future.

Tactic 1. Personal Organization Skills

The BE program is one of the most rigorous found at university, in part, because of the large number of courses a student needs to take; thus, it requires intelligence, motivation, hard work, and a lot of time. The traditional "to do" list many engineering students use for personal organization is flawed, since it doesn't require prioritization and it doesn't assign a timeline. At university, this may be good enough occasionally for just a few days. But, for an engineering student trying to optimize their performance at school, there's a better way.

Let me suggest what I like to call the "clompass" strategy. This is a combination of the words "clock" and "compass." The "compass" part of this strategy deals with assigning priority levels to tasks based on your short-term study duties (e.g., assignments, lab reports, tests, etc.) and also your long-term educational goals (e.g., what grades do you need to achieve to maintain a scholarship? what course do you need to take to get hired by a professor for a summer job? etc.). Then, the "clock" part of this strategy deals with assigning specific time blocks in your daily calendar to only work on the most important tasks, and then assigning final deadlines for those tasks.

Here's how that can work practically. Initially, you would organize your study tasks into 4 categories, such as 1 (important and urgent), 2 (important and not urgent), 3 (unimportant and urgent), and 4 (unimportant and not urgent). Then, schedule specific time blocks in your daily calendar to work only on important study tasks 1 and 2. Finally, for the unimportant study tasks 3 and 4, you either decide never to do them at all because there's not much benefit (e.g., the assignment is only worth 2% of your final grade), or only begin to work on them if and when you are totally finished the important study tasks 1 and 2.

Tactic 2. Learning Styles

There is some scientific evidence that reading is the least effective way for us to learn a task, whereas hearing about the task, seeing the task performed, speaking about the task, doing the task ourselves, and showing others how to do the task are progressively better learning techniques. Whether or not this sequence accurately applies to every single person may be debatable, but the point is that various learning styles may or may not be effective for us.

Now, a good BE program will typically provide you with opportunities to engage in all these learning modalities to some degree, say, through reading the course textbook, listening to lectures, giving verbal project reports to a live audience, doing hands-on experiments, etc. Thus, it may be worth your while to identify your best learning style and strengthen some of your weaker learning styles, so that you can perform better overall in all your courses. I'd suggest finding a book, video, or other resource that can help you do an accurate self-assessment in this regard and then take practical steps to optimize your performance.

So, if you discover that listening is your strongest learning style but reading is your weakest, it may be possible to find an audio version of your course textbook, so you can keep up with the class that way. Similarly, if you find out that doing tasks is your strongest learning style but seeing and listening are your weakest, it may be possible to ask the course professor or teaching assistant to give you extra practice problems to solve on your own beyond those problems that are worked out in the class lectures.

Tactic 3. Study Buddy or Group

Although there will be opportunities for you to work with others on team projects, most of the BE program is geared towards teaching, training, and assessing your performance as an individual. As such, you will spend many hours alone reading textbooks, going over class notes, completing assignments, and doing calculations.

Even so, it will be greatly beneficial for you to find an informal study buddy or group with whom you meet regularly to discuss anything you didn't understand in the lectures or textbooks, show each other how to solve numerical and other problems, divide up research tasks among several individuals who then report back to the group, and even prepare for upcoming tests and exams. Please note, I am simply talking about being part of an informal study or learning group; this is an acceptable practice.

But, I am not encouraging you to copy each other's assignment or reports which the professor specifically asked you to do as an individual without other people's assistance. This is not acceptable, it is dishonest, it amounts to cheating, and it could cause you to fail the course or be expelled from the program. Having said that, another side benefit is that some of these friendships that you establish with an informal study buddy or group could be the start of building your professional network of personal contacts that can last a lifetime and open up many doors of opportunity in the future.

Tactic 4. Speaking and Writing Skills

There's a sometimes well-deserved stereotype of engineers as unable to speak or write in a meaningful way. The reality is that engineers often communicate to the boss, client, customer, funding agency, or each other on a variety of topics, whether it is identifying problems to be solved, presenting the designs of new products, explaining the results of numerical analysis, and so forth. Thus, one of the goals of a BE education is to make sure that students learn how to speak and write effectively. This will require you individually and as part of a team to create audio and video presentations, participate in team discussions, give talks to live audiences, write reports for assignments, projects, and labs, and send emails to fellow students and professors. There is an old proverb I once heard that says something like, "Reading gives you breadth, writing makes you precise, and speaking keeps you ready."

To improve your speaking and writing skills, consider the following ideas: (i) find a good resource, like a mentor, book, video, or audio, on the topic of speaking or writing skills; (ii) visit the university's student center, if it has one, which should have resources in this regard; (iii) ask your professor if they have any sample video presentations or written reports from their past students that can be an example to you; (iv) find a speaker's or writer's club at the university or a similar group in the community which will give you opportunities to get practical feedback as you share samples of your speaking or writing; (v) read engineering research articles or industry reports as an example of proper technical style and formatting; (vi) write down a basic outline in point form of what you need to speak or write about and then start filling in the details; and (vii) practice, practice, practice!

Tactic 5. Performing Well on a QTE

One good study strategy to prepare beforehand for a QTE is this: (i) choose a small number of sample problems (e.g., from the course textbook or from a prior sample QTE) that are representative of the various topics that will be covered; (ii) plan to work on several each day for about 1 or 2 weeks at most before the QTE, so your learning doesn't peak too soon; (iii) seek help from your study buddy or group, the course professor, or the course teaching assistant if you have difficulties; (iv) attend any problem-solving practice sessions offered by the course professor or teaching assistant; and (v) don't do any practice problems the day or night before the QTE, but do something else to give yourself a rest.

One good approach for writing the actual QTE is this: (i) first look over the entire QTE to decide which problems you know how to solve and which might be more difficult, but this will also give you a chance to calm your nerves so you can perform better; (ii) solve all the easy problems to give yourself some confidence and to accumulate points quickly; (iii) then move on to the more difficult problems by solving as much of them as you can to get partial points, but come back to the unanswered parts on a rotating basis because the solutions may come to mind in time; and (iv) if you cannot fully solve certain parts of the difficult problems, then simply write down in words the ideal steps needed to solve those parts of the problem, since this may be worth partial points too.

So, What's the "Take Home" Message of This Letter?

The most important things I want to convey are that a BE university degree will give you the proper and necessary background to work as an engineer, and that there are practical tips that can help you perform optimally in your university engineering studies. It will require time, effort, focus, and creativity on your part, but the opportunities and rewards that a BE degree will open up for your life and career are surely worth it. I'm sure this letter hasn't addressed all your concerns, but I trust it will give you some food for thought.

Kind regards,

R.Z.

Letter 5 Why (and Why Not) to Get a Master's or Doctorate in Engineering

Dear Nick and Natasha,

I trust you're having a successful day and that my letter will bring you some cheer. Especially so, since I hear that you have questions and concerns about what you should do after getting your bachelor of engineering (BE) degree. I know I did. Most people try to find a full-time job right away in university, industry, or government, but others decide to get a master of engineering (ME) or doctorate in engineering (PhD) degree. Some folks who get their PhD go on to do even further specialized training as a post-doctoral fellow (PDF). In any case, all of these are major commitments of time and effort and have their own unique benefits and drawbacks. So, is it worth it? This letter will describe what you might expect from an ME, PhD, or PDF experience, as well as what the pros and cons are for your career in the short- and long-term.

In addition to reading this letter, I recommend that you do a few more things before you apply, if possible. Learn about the reputation, atmosphere, and facilities of the university or technical school to see if it meets your expectations and needs. Discover which professors have openings for an ME, PhD, or PDF candidate. Find out about the teaching focus and research interests of those professors to see who best matches your interests, since one of them will be your supervisor. Make an appointment with those professors to meet them, get a tour of their research labs, and talk with their current ME and PhD students and their PDFs.

Like many of my engineering peers, my personal enjoyment increased in going from my BE to my ME to my PhD to my PDF training, in part, because I was able to focus more and more on a specific research topic I was really interested in. Of course, enjoyment was only one factor that I considered in making my own career choices, since there were other philosophical and practical reasons too. But, of course, this is something you have to decide for yourself. With that said, let's now take a look at some details for a typical ME degree, PhD degree, and PDF training program in engineering (see Figure 5.1).

Master's Degree

Almost all universities that have a BE degree program will also offer an ME degree program in engineering. But, there are a few universities and technical schools that only offer the BE degree, after which students will need to go to a different

DOI: 10.1201/9781003193081-5

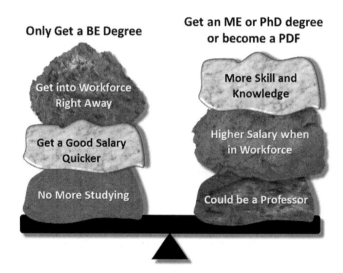

FIGURE 5.1   The pros of getting a BE vs. also getting an ME or PhD or becoming a PDF.

institution to get their ME degree. Other equivalent designations could be MEng (master of engineering), MEngSc (master of engineering science), MASc (master of applied science), or something similar. After completing an ME, you can legally put those letters after your name on your business cards, documents, email signatures, letters, office door, and so forth.

The requirements for a typical ME include several courses, plus a major research thesis for which the student engages in analytical/mathematical modeling, computational analysis, experimental testing, and/or on-site field observations on a unique topic, as well as the occasional back-of-the-envelope calculation. The findings of the thesis are often expected to make a new contribution to engineering by 1 article published in a peer-reviewed journal or 1 presentation given at a conference. This is commonly a 2-year program, the student will be paid a modest salary, and they will often be a teaching assistant for BE courses that will involve facilitating tutorial sessions, grading assignments, and/or running hands-on lab exercises.

Some universities offer a shorter version of the ME program that requires a greater number of courses, but only a minor research project that is simply intended to demonstrate the student's basic competence, that may not make a unique contribution to engineering that is publishable as an article. This is commonly a 1- to 2-year program, the student will not be paid a salary, and they may or may not be asked to be a teaching assistant.

For both versions of the ME program, the student will have a professor who will be their supervisor to guide them in their studies, although an additional co-supervisor may be involved especially if they bring some knowledge, skills, or resources that the primary supervisor lacks. Also, if the student knows for certain that they will stop after getting an ME, then it won't matter to potential employers whether or not it was from the same institution where the BE was obtained.

The benefits of an ME vs. only a BE are (i) higher salary in the workforce, (ii) more skill and knowledge that often increases the chances of employment, (iii) the option of being a professor at a university or technical school that offers only BE degrees, and (iv) the option of doing a doctoral degree. The drawbacks of an ME vs. only a BE are (i) a 1- to 2-year delay of getting into the workforce and collecting a good salary compared to peers, (ii) some employers only want BE graduates because they are younger and can be more easily trained to meet the company's objectives, and (iii) more studying to do.

Doctoral Degree

Virtually all universities that have an ME degree program will also offer a PhD degree program in engineering. But, there might be a few universities and technical schools that only offer the ME degree, after which students must go to a different institution to get their PhD degree. Another equivalent designation could be the DSc (doctor of science), or something similar. After finishing a PhD, you can legally put those letters after your name on your office door, business cards, letters, documents, email signatures, and so on.

The requirements for a typical PhD may or may not include several courses, but it will always include a major research thesis for which the student will engage in analytical/mathematical modeling, computational analysis, experimental testing, and/or on-site field observations on a particular topic, plus some back-of-the-envelope calculations from time to time. The findings of the thesis must make a major contribution to engineering by a series of articles published in peer-reviewed journals and presentations given at conferences. This is commonly a 3- to 5-year program, the student will be paid a modest salary, and they will be a teaching assistant for BE or ME courses that will involve facilitating tutorial sessions, grading assignments, and/or running hands-on lab exercises.

The PhD student will have a professor who will supervise them in their studies and provide full or partial salary support, although an additional co-supervisor could be involved if they bring some knowledge, skills, or resources that the primary supervisor lacks. Also, because the program is so long, the student may need to apply for government grants, industry sponsorships, or other scholarships to supplement any salary funding their supervising professor is providing.

Some more rigorous PhD programs also require the student to complete a major series of exams (separate from any course exams) in order to test their knowledge on certain core topics in their engineering discipline, as well as passing a preliminary exam on their proposed research topic before they get too far into their research work. The student's performance on these exams will be judged by a committee of university and/or industry experts.

If someone gets a BE and ME from the same institution, it is probably advisable to get their PhD somewhere else to give them a broader background, so they are not perceived by potential university, industry, or government employers as being too narrow intellectually and experientially. But, if they did the BE and ME at different institutions, then it's okay to go back to one of those to do PhD studies, since they will still be perceived as having a diverse educational background. Moreover, many people like to work in industry or government for a few years to earn some money and gain experience, and then come back to university later to do their PhD, but this approach is not critical for a long-term successful career. Also, some universities let students complete only the first year of an ME and then, if their course grades are excellent, to transfer directly to a PhD program, which saves a year.

The benefits of a PhD vs. only an ME are (i) higher salary in the workforce, (ii) more skill and knowledge that often increases the chances of employment, (iii) the option of being a professor at a university or technical school, and (iv) the option of doing a post-doctoral fellowship. The drawbacks of a PhD vs. only an ME are (i) a 3- to 5-year delay of getting into the workforce and collecting a good salary compared to your peers, (ii) some employers only want ME graduates because they are younger and can be more easily trained to achieve the company's goals, and (iii) more studying to do.

Post-Doctoral Fellow

All universities that offer a PhD in engineering will also provide opportunities for a PhD graduate to do extra training as a PDF. The trainee is also informally called a "post doc," and the training itself is called a "fellowship." This training experience is not a degree program, so there are no additional designatory letters to legally place after your name, unlike a BE, ME, or PhD degree.

A PDF has no courses or research thesis to complete. Instead, the goal is to give the PDF extra training in research under the supervision of an experienced professor. The PDF will often informally supervise ME and PhD students doing their own research theses and be involved in applying for government grants, industry sponsorships, or other scholarships to fund the ME and PhD students. The research may include analytical/mathematical modeling, computational analysis, experimental testing, and/or on-site field observations, as well as some back-of-the-envelope calculations when needed. The research must make a major contribution to engineering by a series of scholarly articles published in peer-reviewed journals and presentations given at conferences. The PDF is usually under contract for a 1- to 2-year period, they will be paid a much higher salary than they had during their PhD studies, and they may be teaching BE, ME, or PhD courses and/or acting as the teaching assistant for such courses.

If someone gets a BE, ME, and PhD from the same university, it is probably advisable for them to become a PDF somewhere else to gain a broader background, so they are not looked upon by potential university, industry, or government employers as being too narrow intellectually and experientially. But, if they did the BE, ME, and PhD at various universities, then it's okay to go back to one of those as a PDF, since they will still be looked upon as having a diverse educational background. As mentioned earlier, many folks like to work in industry or government for several years to earn money and obtain experience, and then come back later to university to be a PDF, but this strategy is not critical for long-term success.

The primary benefit of a PDF position is that it gives you more skill and knowledge that increases your chances of employment as a professor at a university or technical school. The primary drawback of a PDF position is that there is a 1- to 2-year delay of getting into the university, industry, or government workforce and collecting a much better salary compared to your peers.

Will I Be Too Specialized?

You might be concerned that these highly specialized engineering degrees and experiences may lead to a dead end that forever locks you into a career that you later don't really like or that doesn't meet your financial or other goals. Please be assured that this is not necessarily the case, so don't feel paralyzed in making these decisions. There are always potential ways to change directions in your career at a later time, especially if the opportunity clearly arises.

In my case, during my BE studies in mechanical engineering I chose to do 2 courses and my graduating thesis project in biomedical applications. Later, I did my ME in mechanical engineering with a research thesis in the area of aerodynamics or wind engineering. I investigated how wind flows around big overhead electrical power lines when they accumulate ice on their surface during winter. The ice actually forms into a teardrop or pear shape, so it acts like a small airplane wing. And once the wind starts to blow, the power lines can start to move up and down so violently that it damages the power lines and causes major electrical shortages. I did all sorts of interesting wind tunnel tests like using smoke visualization techniques and so on. I enjoyed my experience so much that I considered staying in this area for the rest of my career.

So, when I finished my ME, I applied to several PhD mechanical engineering programs where I could pursue my interests in wind flow and flow induced vibrations. But, I also applied to several PhD mechanical engineering programs that focused on biomedical applications. I was torn between these 2 topics and just wanted to see which doors opened and closed. The best option that finally presented itself for practical and financial reasons was one of the PhD programs in biomedical applications. My PhD thesis research topic was to use medical ultrasound technology to analyze the mechanical stresses experienced by total knee implants. I also became a PDF for 2 years in the same area. And I happily stayed in this area for the remainder of my career.

But also consider some of the colleagues I've personally known who made even bigger shifts. Some of them did their ME or PhD degrees in engineering, yet then decided they really wanted to be an MD (doctor of medicine); they became doctors with thriving full-time medical practices in reputable hospitals. Similarly, I know another colleague who did an ME, but then completed a master's degree in intellectual property law and a master's degree in business administration; this person went on to have satisfying full-time work in the upper levels of administration in a university as a science and technology patent officer. I also knew someone who completed a BE degree, but then several years later decided to go to law school, after which they had a successful career as a lawyer.

So, What's the "Take Home" Message of This Letter?

The main goal of this letter was to give you insights into some of the most important practical aspects of getting an ME or PhD degree, as well as being a PDF. Also, I'm sure that you've noticed that there are pros and cons to each of them, which you can consider seriously in charting the course of your engineering career. Regardless of how your career unfolds, I wish you the best.

All for now,

R.Z.

Letter 6 Professional Engineering Status: Societies, Certifications, Designations

Dear Natasha and Nick,

Greetings to you from my office. I'm looking at the various engineering books that surround me. Let's assume in an ideal world that I know everything in them perfectly. And that I even have the highest possible academic training with a doctoral degree in engineering. Those facts would still not always legally qualify me to be an engineer who can authorize the fabrication of devices, the construction of structures, etc., especially when public health and safety are affected.

So, I'd still need a professional engineering society (PES) that's legally authorized by the government to certify and designate me as a professional engineer (PE), or something similar. And, that's what this letter is about, namely, the functions, benefits, and drawbacks of professional engineering societies, certifications, and designations. Now, keep in mind that the details of what I write below may vary from country to country and PES to PES, but this letter should still give you an accurate appreciation of this topic. With that stated, let's get to it.

What's the History of the PES?

During ancient and medieval times, groups of skilled craftsmen or tradesmen—the forerunners of today's engineers—would often organize themselves into trade guilds or associations. Sometimes, the guilds acted like mutual support groups that provided moral and material assistance to its members. At other times, the guilds were more like political lobby groups or labor unions who advocated with the government for the rights and rewards of its members. In other instances, the guilds were exclusive clubs that focused on the job security of members by ensuring that non-members could not legally offer similar competing services to the public.

Much later in France, formal professional societies were created for military engineers in 1690 and civil engineers in 1716, which was followed by the establishment of a training school for civil engineers in 1747. In Britain, an informal Society of Engineers was established in 1771 that met over meals to discuss solutions to various engineering problems. Then, in the aftermath of the Industrial

DOI: 10.1201/9781003193081-6

Revolution (1769–1829), formal professional societies were organized in Britain for civil engineers (1828), mechanical engineers (1847), electrical engineers (1871), and, sometime even later, chemical engineers (1922).

Then, as reported by S.H. Christensen and colleagues in their book *Engineering Identities, Epistemologies and Values* (Springer, 2015) an important meeting occurred in 1960 called the Conference of Engineering Societies of Western Europe and the United States of America. The conference declared that "a professional engineer is competent by virtue of his/her fundamental education and training to apply the scientific method and outlook to the analysis and solution of engineering problems...In due time, he/she will be able to give authoritative technical advice and to assume responsibility for the direction of important tasks in his/her branch." (p. 157–158). The legacy of these early trade guilds and engineering societies carries down to our own day.

What's a Modern-day PES?

Today's PES is a formal organization that is recognized by a regional or national government and, thus, has the authority to confer, withhold, or remove the status of PE to an engineer employed in university, industry, or government (see Figure 6.1). Don't confuse a PES with the many other scientific or engineering organizations that an engineer can also join that may be related to their work, but these groups do not have any legal authority to confer PE status on anyone.

A PES mainly exists to give *formal certification* to members who have the proper education, experience, competence, and ethics to do engineering work that is high quality, reliable, safe, and honest. Senior or supervising engineers in industry or government, as well as professors of engineering at the university, are often required by their employers to be formally certified. But, others get certified because it can be a strategic career move in the long-term, not because they are required to do so.

In any case, certified members then may be given official *designatory letters* they can add after their name to visibly assure the government and the public they are qualified experts who can do the job. These are the letters PE that I've already been using here, or it may instead be PEng (professional engineer), CEng (chartered engineer), or something similar, to indicate professional certification. And, it may or may not include additional letters like A (affiliate), M (member), or F (fellow), or something similar, that are attached to the acronym for the official name of the PES to indicate membership.

So, for example, let's consider an engineer who has the post-name designatory letters PhD, CEng, MIMechE. The PhD signifies their university education as a doctor of engineering, the CEng signifies their official certification as a chartered engineer, and the MIMechE signifies they are a member of the Institution of Mechanical Engineers. Or, let's assume an engineer has the post-name designatory letters BASc, PE. The BASc signifies their university education as a bachelor of applied science and the PE signifies their official certification as a professional engineer, but in this case the particular PES does not offer designatory letters to indicate the engineer's membership in the PES.

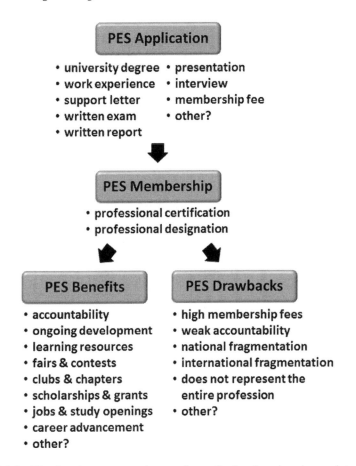

FIGURE 6.1   The functions, pros, and cons of a professional engineering society (PES).

Why and How Do You Join a PES?

In many cases, employers require PE status for their senior or supervising engineers in industry or government, or professors of engineering who teach and do research at a university. Obviously, these folks have no choice, since getting PE status is a condition of their employment. In other cases, junior engineers, research engineers, and others, may not be required by their employers to do so, but they decide to get certified anyway because it can help them increase their salary, promote them to a higher position, improve their resumé for future job applications, etc.

If you need or want PE status, you may need to satisfy one or more of these requirements. You'll definitely need a bachelor's (BE), master's (ME), or doctoral (PhD) degree in engineering—with these or other equivalent designatory letters—from a university or technical school. You may need one to several years of practical work experience. An engineer with PE status who's worked with you may need to vouch for your competence and honesty via a support letter, email, etc. There could be an exam in engineering ethics, engineering law, etc., that needs to be passed. You may be asked to submit a written report or give an in-person or online presentation summarizing your career-to-date to show your engineering ability, as well as describing your long-term career goals to show your commitment to lifelong learning. You might be interviewed by a committee from the PES. There's probably a one-time application fee and/or an ongoing membership fee. And there could even be other requirements that I haven't listed here.

Now, if you don't require PE status and have no great career ambitions, then carry on as you are without giving it a second thought. But, if you need or want PE status, then it's probably best to keep it active for the rest of your career because of the advantages discussed earlier. If you resign or let your PE status expire, it may not always be quick or easy to get back again, if a circumstance arises in which you need it.

What Are the Benefits of a PES?

In addition to giving an engineer a legally recognized PE certification and designation, a typical PES may offer a variety of extra advantages and opportunities to its members. I list some of the key ones, but there may be others.

A PES can *provide accountability*. Membership in a PES is a gentle encouragement and, at times, a stark reminder to the engineer about their responsibility to do high-quality, reliable, safe, and honest work. If a PE is accused of being incompetent or unethical, a PES can temporarily suspend the PE status. If the engineer remains employed, they will likely need to work under the direct supervision of another PE. If the accusations of dishonesty or incompetence are eventually proved false and/or the engineer has undergone sufficient retraining, their PE status may be restored. If not, the PE status can be permanently removed.

A PES may encourage *ongoing development*. Just because an engineer has PE status, of course, that doesn't mean that they have nothing new to learn through in-person or online events and resources. There are always new scientific discoveries, emerging technologies, changing government regulations, personal organization skills, leadership skills, communication skills, entrepreneurial skills, and so forth, that they can learn about in order to make them more efficient and effective in their work.

A PES might provide *learning resources*. This allows members to have free or low-cost access to physical or digital scholarly journals, trade magazines, and newsletters, so members can keep up with the latest scientific findings, technological changes, industry developments, and government policies. Access to university textbooks, reference works, biographies, histories, etc., can provide foundational technical and other information. And, online software can help members with 3D drawing, computational analysis, etc.

A PES will sometimes organize *science and technology fairs and contests*. Student and full members of the PES can give short presentations, create physical displays that highlight their work, or enter their own inventions in competitions against other participants. These events are a way of reaching out to the general public, gaining the interest of potential private donors and industry sponsors, allowing attendees to network with each other, and recruiting new members to the PES.

A PES might sponsor *local clubs and chapters*. These could be focused on student members who are still studying at university, or working engineers in university, industry, or government. Either way, clubs and chapters provide great opportunities for growing an engineer's personal network of contacts that can open up doors of opportunity in the future. Such clubs and chapters often host events with guest speakers, discussions, fundraising, meals, and so forth.

A PES could award *scholarships and grants*. This can help financially support student members to complete their BE degrees at university and full members who are pursuing an ME or PhD degree in engineering. Moreover, funding can greatly aid members who are working on research projects in university, industry, or government. And, such funds can be used to cover the travel expenses of members going to academic, industry, or government conferences.

A PES can advertize *new jobs and study opportunities*. This mainly occurs through emails, physical or digital newsletters, or postings on their website. Such information is critical for various people, such as new engineering graduates seeking full-time employment. Or, it could be relevant to those who've just graduated with their initial BE degree, as well as working engineers, who now wish to pursue an ME or PhD degree in engineering. Or, it may be useful for working engineers who are looking for a new job.

A PES may help with *career advancement*. Junior engineers just entering the workforce, engineers engaged mainly in pure research, and others, are often not required by law or their employers to obtain PE status. And so, many engineers in these situations choose not to join a PES or get PE certification because it doesn't benefit them. However, there are others in the same situation who still choose to join a PES so they can be a certified PE. The reason is that it opens up new possibilities, such as a salary increase, a job promotion to a more senior position, a better resumé that is more attractive to future employers, etc.

What Are the Drawbacks of a PES?

A PES, like any organization, can have shortcomings. There is an adage that says something like, "The tail wags the dog." This means that someone or something that is originally meant only to offer assistance, if one is not careful, can eventually take over total control and seek only its own benefits. So, although a PES is officially meant to certify, designate, support, and oversee its engineering members, it can potentially become a rigid and decayed system that has forgotten its original purpose. Below, I discuss some of the most important drawbacks, but there could be others.

Members typically have to pay rather *high membership fees*. This is often required to obtain and/or maintain PE status, but sometimes these fees are too high for the few benefits that members actually get in return. To be fair, some employers are willing to reimburse their engineers for these costs. Even so, that's why some engineers who don't really need to be a PE for their particular work may refuse to ever get one. Other engineers will criticize the PES as simply being a scam to get money to keep itself running, although it doesn't really have any true value to them on a day-to-day basis.

Another common problem is *weak accountability*. A PES is supposed to keep its members accountable to ensure they are doing their duties ethically and giving products and services to customers that are high-caliber, reliable, and safe. Sadly, a typical PES may have absolutely no idea what its own members are doing, since it doesn't require a regular progress report from its members via in-person interview, written report, email, telephone, video, and so forth. And, conversely, a PES may not even send any regular communications to its members, other than a reminder to pay the annual membership fees.

On top of that, there's the challenge of *national fragmentation*. This means that some countries, especially large ones, have a different PES for each region of their country. The reason for this is the existence of slightly different laws from region to region. Even so, the government and each PES are creating barriers for a PE to work in different regions of their own country. The PE may have to rewrite a certification exam or redo an interview each time they move, which is often a waste of time, energy, and money. So, whenever possible, in my view, it's better to have one single PES—either for each engineering specialty or combined together—that manages all engineers in the whole country.

Similarly, there's the issue of *international fragmentation*. This is more understandable, since each country has its own distinct laws that the PES must follow. But, there's a great irony here. One country might have a PES with only loose requirements for conferring PE status to its engineers, yet it won't officially accept the validity of PE status from another country whose PES has much higher requirements. Also, a PES may totally reject an immigrant engineer and ask them to go back to university to get another engineering degree. Such re-education is a huge barrier that may be too large for some people either financially or psychologically, which means another potentially talented engineer is being lost. So, in my view, it's better to have a more realistic pathway for an immigrant engineer to transfer their PE status to the new country. For instance, the immigrant engineer could demonstrate their professional competency by passing a few equivalency exams and/or interviews, as well as getting some practical experience.

Finally, a PES *doesn't fully represent* the engineering profession. As mentioned already, a senior or supervising engineer or a professor of engineering at a university is often required by their employers to be members of a PES and obtain PE status. Yet, there are others who don't need PE status, but get certified anyway because they think it's strategic for their long-term career plans. But, of course, this doesn't represent all the many engineers that are not required to do so and, thus, don't obtain PE status. Their ideas and dreams are not represented by a PES, and their ethics and competence are not monitored by a PES. This can turn out to be a big problem in the far future for the profession of engineering as a whole.

So, What's the "Take Home" Message of This Letter?

My intention was not to persuade you one way or another about joining a PES and getting PE status. So, I've done my level best to outline the main pros and cons. I'll remind you again that what I've written will not exactly apply to every context, since it can vary from one country to another and from one PES to another. Having said that, I'll bid you farewell for now, having every confidence that you'll make the right decisions for your engineering career.

Seize the day,

R.Z.

Part III

Principles

Letter 7 The Science of Engineering

Dear Nick and Natasha,

Salutations! I'd like to start this letter with my version of an amusing tale I heard long ago, which I hope you enjoy. One day a mathematician said to a bystander, "Look at that freshly fallen apple sitting under that tree. I'll bet you a big bag of money that you can't eat the apple." The bystander asked, "What's the problem? I'll just walk over there right now." "Ah," the mathematician said. "Wait. There's only one rule. Each time you move, it must be exactly half the distance remaining to the apple. You see, as a mathematician, I know theoretically that you'll never reach the apple. And I'll win the bet."

The bystander—with a slightly mischievous twinkle in his eye—quickly agreed. With each move, he got closer to the apple by exactly half the total initial distance. First he moved 1/2 of the total initial distance, then 1/4 of the total initial distance, then 1/8 of the total initial distance, and so on, until he was standing in front of the apple. The bystander stooped down, picked up the apple, took a bite, walked back to the mathematician, and took his big bag of money.

"Objection!" bellowed the mathematician. "You must have broken the rule." The bystander said, "No, I obeyed the rule. But, there's one thing you didn't know about me. You see, I'm an engineer. I solve real-world problems using practical solutions. Although the apple could never theoretically be reached by a mathematician like you, it was practically close enough for an engineer like me!" And the engineer, now rich and eating the apple, went on his merry way, as the stunned mathematician looked on.

Let me assure you that I've got no ill feelings towards mathematicians or other scientists. My point in telling your this tale—and the point of this letter—is that fruitful engineers clearly understand how to skillfully use basic scientific principles, procedures, and tools to accomplish their goals. This helps them deliver effective solutions for all types of practical problems. And that's what I mean by the science of engineering. Therefore, I'd like to discuss some very foundational concepts necessary for the "scientific" engineer that you may or may not be familiar with. Let's get into the details.

The Lifecycle of Engineering Products

The main purpose of engineering regardless of specialty is to provide high-quality products to clients, customers, and society at large that can solve practical dilemmas. These products could be anything from small gadgets to midsize machines to huge structures, or even unseen processes, software packages, and digital information.

DOI: 10.1201/9781003193081-7

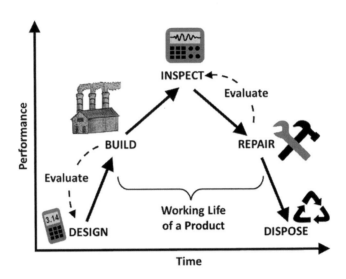

FIGURE 7.1   The design-to-dispose lifecycle of a product.

Whatever the case, all products have a design-to-dispose lifecycle that, in some ways, parallels the birth-to-death lifespan of organic matter, plants, animals, and humans. The lifecycle of a product can be split into different phases in different ways, but I like to think of it as having 5 key phases (see Figure 7.1). Now, I'm not going to get into the details of each phase in this letter because they each could and do have entire books and university courses devoted to them. But, I think it's worth at least highlighting what they are in case you're not familiar with them, as a way of peaking your interest.

The *design* phase means conceiving small or large improvements on the form or function of older products, as well as devising completely new ones. The *build* phase means using state-of-the-art machines, methodologies, and materials to fabricate both initial prototypes and the final products. The *inspect* phase means evaluating the quality and performance of products and making all necessary recommendations or changes. The *repair* phase means fixing or upgrading products that are not functioning properly so that they meet the desired safety standards, technical specifications, user needs, and so on. The *dispose* phase means storing, discarding, disassembling, or recycling products that are irreparable or out-of-date in a safe, legal, ethical, and effective way.

The actual *working life* of a product can be thought of as starting just after it is built to just after the time when it can no longer be repaired or upgraded effectively. You'll also notice that there is a back-and-forth iterative process between the *design* and *build* phases until the final version of the product has been decided and then created for final use. Also, there is a back-and-forth iterative process between the *inspect* and *repair* phases over a long period of time until the product becomes irreparable or out-of-date.

Engineers are often involved in physically carrying out, or at least planning and managing, every phase of a product's lifecycle. In some cases, it's possible that a particular individual engineer is going to do this, but there also may be several different engineers involved in different phases. Also, once a product is in use by the client or customer, it may be left to others like technicians to physically be responsible for repairing and eventually disposing a product. Whatever the case, engineers can accomplish their goals optimally in each phase of a product's life-cycle by using the scientific method combined with the "5 golden tools" in their toolbox, as described below.

The Scientific Method for Engineers

The English word "science" comes from the Latin word "scientia," which simply means "knowledge" of any kind. But, the word has come to be almost exclusively associated with having knowledge about, and inquiring into, the fundamental properties and processes of the material universe. And so, you could say that "pure science" is about discovering new things about biology, chemistry, physics, and mathematics, while "applied science" refers to using those same discoveries to solve practical problems in society. In our case, as applied scientists or engineers, we want to use science to create devices, harness natural energy sources, offer services, optimize processes, etc., that make a tangible difference in people's lives.

But, the line separating pure science and engineering is blurry. To reach our engineering goals, we often first need to do pure science research by identifying the problem we want to solve (e.g., osteoarthritis of the human hip) and then evaluating the attributes of the possible solutions (e.g., is this material strong enough to build a successful artificial hip implant to treat osteoarthritis of the human hip?). This type of inquiry, in essence, is doing science.

However, there is a widely-established procedure for doing this called the "scientific method," which, at its core, is this: make an initial hypothesis, perform a test, and revise the initial hypothesis, if needed, based on the test results. This is done repeatedly to get a progressively more accurate understanding of the phenomenon being researched. The ancient Greek philosopher Aristotle (384–322 BC) promoted observation, experimentation, and systematic data classification and, thus, is often considered the founder of the scientific method. But, the Scientific Revolution of the 1500s and 1600s more securely laid the foundation for modern science.

In day-to-day practice, as an engineer doing pure science research, you will need to do a lot more than the core of the scientific method suggests. In the real world, the actual procedure for a single research project might look more like the 12-step list below. But, keep in mind that steps 2 to 7 can have a slightly different order, some of them may need to be repeated or revised as you go along, and sometimes a few of them can even occur simultaneously in parallel. A lot of this will really depend on the specific circumstances of the research project, the engineering team, or the individual engineer. In any case, the following steps are worth considering seriously as you plan your project.

1. Define the problem (e.g., research questions, client needs).
2. Identify the tasks (e.g., team roles, delegate, outsource).
3. Decide the timeline (e.g., Gantt chart with mini deadlines).

4. Review the literature (e.g., journal articles, conference talks).
5. Design the methodology (e.g., computations, experiments).
6. Gather the resources (e.g., funds, equipment, specimens).
7. Make a hypothesis (e.g., initial predictive guess about results).
8. Perform the investigation (e.g., computations, experiments).
9. Analyze the data (e.g., statistical comparisons, best-fit curves).
10. Interpret the results (e.g., practical implications, limits, trends).
11. Revise the original hypothesis (e.g., test results are unexpected).
12. Communicate the findings (e.g., journal articles, conference talks).

However, I should mention here that for a very narrow research problem, question, or topic, a single research project or small series of research projects may be enough to solve the problem conclusively. Yet, for a very broad problem, question, or topic, this may require the efforts of numerous researchers over a long period of time. In that case, as more and more projects are done exploring various facets of the problem, question, or topic, then a fuller understanding is progressively gained. Eventually, the accumulation of fresh new knowledge can start to reach a plateau and different researchers may start to confirm the same results over and over again. At this point, we may have arrived at a reliable explanatory conceptual model (i.e., paradigm, theory, or law) that best fits a wide range of established facts for the problem, question, or topic. Such models will remain in place for a long time as guideposts for all future research, although they can undergo small refinements along the way. But, any major changes will usually only be accepted by the research community if there are compellingly persuasive new results.

And, finally, the above 12-step approach for pure science research can also be used for project planning for any kind of engineering endeavor at any point in a product's lifecycle. At first glance, the 12-step approach may seem most suitable for the design and build phases of a product's lifecycle when it is first being researched and developed (hence, the term R&D). And that's what a lot of R&D engineers do. But, an engineer can also use it when inspecting, repairing, or disposing a product with only minor differences in the details depending on the situation. So, for example, an engineer might do step 4 by consulting an industry handbook or product manual for how to safely dispose a product, rather than reading a scholarly journal article. Steps 7 and 11 might require an engineer to perform on-the-spot diagnostics plus informal guesses to determine the reasons why a product isn't working properly, rather than making formal hypotheses. Step 12 may mean an engineer writes a short post-task report to their boss about how things went, rather than composing a lengthy scholarly journal article. And so on.

The "5 Golden Tools" in the Engineer's Toolbox

An engineer's work can involve designing, building, inspecting, repairing, or disposing a product, or it can be focused on pure science research. To do this successfully, engineers should have certain essential tools in their toolbox. I like to call these the "5 golden tools." You may or may not have much experience in using these tools, depending on where you are in your education or career. And so, I'd like to describe each of them briefly, tell you a little about their potential pros and cons, and also give you a few tips along the way.

The first tool is *mathematical (or analytical) modeling*. This tool involves using first principles and fundamental concepts to develop sophisticated equations that are meant to accurately and thoroughly describe physical phenomena. This methodology is especially useful in pure or theoretical engineering research for a master's thesis, doctoral thesis, or post-doctoral fellowship. This kind of work requires an advanced grasp of mathematics and physics that the average engineer may never see or use in their day-to-day work or an entire career. These equations are sometimes extremely complex with many constants, variables, differentials, integrals, matrices, nested brackets, non-linearities, etc. But, because they can sometimes be impractical to use, these equations can be trimmed down using simplifying assumptions to arrive at practical easy-to-use formulas for everyday engineering work. It is these simpler formulas that student engineers primarily learn from their textbooks and industry engineers mainly use from their handbooks. The formulas are seen as "good enough" approximations of the actual equations, but strictly speaking, they are limited and incorrect.

The second tool is *computational analysis*. This tool involves using ready-made commercially-available computer software packages to analyze various engineering problems in electromagnetism, fluid flow, heat transfer, mechanical stress, process optimization, etc. These software packages simulate a 1-, 2-, or 3-dimensional object or process of interest using a chain-like series or mesh-like structure of numerous small units called "finite elements." Each finite element's physical behavior is represented by a mathematical equation. The user sets up the initial conditions of the problem to be solved, inputs the known numerical values or formulas, allows the software to run the analysis, and then receives the outputted solutions. The clear advantage of these software packages is that they save time, effort, and stress for the student or working engineer, who don't have to solve complicated equations by hand as done in the "old days." The other benefit is that software packages can explore conditions that could be difficult, dangerous, or expensive to do experimentally in the lab or by on-site field observations. The temptation is to rely too much on the software package, but without comprehending the underlying physics of the problem or the equations the software uses. Of course, there is also the issue of the reliability of the results. So, I highly recommend validating your computational analysis by comparing the results against those from experiments. You may only need to do this a few times if you use the computational model to solve very similar problems all the time. But, if you work on a variety of different problems, then you may need to do experimental validation each time. How close computational results must be to experimental results for them to be considered valid is debatable, a matter of judgment, and may depend on the situation. And, finally, keep in mind that there are many other types of software packages that you can purchase, so you can more easily perform statistical analysis, image analysis, technical drawing, artistic drawing, etc.

The third tool is *experimental testing*. This tool involves doing tests on physical specimens or prototypes in a controlled laboratory setting to see how they respond to different conditions. To do so, the researcher has to mobilize physical resources (e.g., specimens, prototypes, measuring equipment, etc.) and trained personnel (e.g., student engineer, research engineer, technician, etc.) to perform the tests, collect, analyze, and interpret the data, and report the findings. The obvious benefit, if the studies are done properly, is that results are often considered to be self-evidently reliable because physical specimens and prototypes are used and realistic conditions are based on well-established research protocols. For engineering, these research protocols can be obtained from the ASTM (American Society for Testing and Materials) and the ISO (International Organization for Standardization), but if they don't address your particular need then journal articles, textbooks, or handbooks may provide some guidelines. A potential disadvantage is that laboratory-based experiments still often make some simplifications of their own (e.g., specimen size is adjusted, conditions are idealized, etc.) in order to make it feasible to perform the experiments. Yet, this may not always exactly replicate the real-world conditions under which the devices, machines, structures, etc., being simulated really function. Also, experiments often involve measurement equipment that's used together with specialized computer software to record and process data, but this setup has to first be adjusted or calibrated using tests that produce known results to ensure the measurements are accurate and repeatable. Another major drawback to experiments is the often high financial cost to purchase or rent all the physical resources used and to pay the salaries of all the personnel.

The fourth tool is *on-site field observations*. This tool involves making observations, doing preliminary calculations, and taking measurements on an engineering product on-site (i.e., "in the field") where it is actually being used by a client, customer, or society. So, for example, the product could be a machine operating in a factory, a bridge that conveys cars and trucks over a river, a computer software package being used by an organization, and so on. This is the kind of work that maintenance engineers routinely do, but student engineers might do this too as part of a master's or doctoral research project. The results obtained from on-site field observations have the great advantage of being realistic, since the product is in its intended environment of use and is subjected to conditions for which it was originally designed. Such information is absolutely vital for engineers to know, so that public health and safety is not endangered and so that they know how to make future improvements to the product to make it more efficient. But, there are some drawbacks too. For instance, it's often difficult to properly investigate a product's performance in the field because sometimes you can't change any input parameters step-by-step to see their effects, as you could in a controlled laboratory setting (e.g., going from low to medium to high wind speeds blowing over a model of a bridge being tested in a wind tunnel). You just have to accept the conditions in the field as they are at that moment (e.g., unpredictable wind speeds blowing over a full-size bridge). Also, you may not always have easy access to the product if the company doesn't want outsiders in their factories to protect certain information from leaking to their competitors in the marketplace, if the government only allows its own employed engineers to have access to the public works it has constructed, etc.

The fifth tool is *back-of-the-envelope calculations*. This tool involves using first principles (i.e., foundational theories) to do a simple and quick calculation to get a rough estimate of a numerical result. In the "old days," engineers would use a pencil, paper (even the back of an envelope, if it was handy), and pocket calculator to make a calculation. But, today engineers can also use their smartphone or laptop computer. Although this tool isn't meant to replace a more precise, accurate, and thorough engineering analysis using the other tools described above, it can be very useful to engineers in different contexts. Student engineers routinely use this approach to solve problems the professor gives them on assignments and exams. Research engineers can apply this tactic to double-check if their more detailed computational analyses and experimental tests are giving reasonable results and then make any needed adjustments. Industry engineers can employ this method to do an initial assessment of how well a device, machine, structure, etc., is functioning and then make practical recommendations or changes. It's not just about plugging numbers into equations or formulas. Rather, I recommend doing a back-of-the-envelope calculation in a step-by-step way, as follows:

1. Define the problem or question.
2. Draw a sketch to visualize the problem.
3. Write down all the known quantities.
4. Write down all the unknown variables.
5. Write down any assumptions to simplify the problem.
6. Write down all relevant equations or formulas.
7. Write down any "safety factors" to be applied.
8. Solve to get numerical results for the unknown variables.
9. Reflect on whether the results make physical sense.
10. Recheck each step, if step 9 suggests there's an error.

However, I should point out here that unthinkingly relying on equations or formulas is dangerous, since engineers need to understand how those equations or formulas represent real physical phenomena. And memorizing equations or formulas isn't very important, since you can always find them in textbooks and handbooks. Personally, I would even recommend that you try to use first principles to mathematically derive some of the equations or formulas you frequently use in your engineering studies or work. I have done this many times. I can tell you that it really solidified in my mind exactly what those equations or formulas meant. And it has helped me to better understand the physical phenomena they represent.

So, What's the "Take Home" Message of This Letter?

In essence, I want to encourage you to learn how to skillfully use basic scientific principles, procedures, and tools to deliver practical solutions to address the real-life problems of society. It's thrilling for me to know that you, as an engineer, can then go on to create a bridge or building, a prosthesis or power station, or a robot or rocket. This gives me hope for the future of our engineering profession and for human civilization.

Best regards,

R.Z.

Letter 8 The Art of Engineering

Dear Natasha and Nick,

I send you good wishes from my office which, to be honest, is perhaps a bit too dull. In contrast, some folks have interesting paintings, photos, posters, or statues in their office to inspire them in their work. I've thought about getting one of those diagrams or sculptures of the human brain that is divided into a left and right side. Based on some evidence from neuroscience and psychology, the left side supposedly deals with logic, order, rationality, step-by-step processes, and so forth. And the right side supposedly deals with creativity, emotions, imagination, lateral thinking, and the like. Together they make up a whole brain that allows us to function effectively in this world.

So, why should this be any different for engineers? Sometimes I think there's a misconception that we engineers have been overly educated and trained to use only the left side of our brains, while completely ignoring or at least minimizing the importance of the right side. There may be some truth to that. But, any engineer who has a successful and satisfying career will probably tell you that they use the right side of their brain more than they often realize. It's an integral part of who engineers are and what they do.

Engineers can be artistic by thinking of truly innovative ways for how technology can bring about a better future for human civilization in the short and long term. Engineers can use their imaginations to deliver products and services that positively affect human psychology and that also appeal to the 5 physical senses. Engineers can embrace curiosity, mistakes, and play as creative and valid pathways to discovery, design, and even problem-solving. And, that's what this letter is about—the art of engineering.

W.I.B. Beveridge (1908–2006) wrote in his classic book *The Art of Scientific Investigation* (Vintage Books, 1950/1957) that "Imagination is of great importance not only in leading us to new facts, but also in stimulating us to new efforts, for it enables us to see visions of their possible consequences. Facts and ideas are dead in themselves and it is the imagination that gives life to them." (p. 78). So, what I'd like to do here is discuss some of my favorite ways we engineers can stir up and utilize our creativity (see Figure 8.1), although I'm sure you'd be able to find other good resources written by real experts on the psychology of creative thinking.

Ask Questions

First off, don't be afraid to have an inquiring mind. Ask all kinds of questions. Odd questions. Silly questions. Funny questions. Bizarre questions. Persistent questions. If you don't ask questions, you'll never get answers. There's an old proverb that says, "There's no such thing as a stupid question." Generally speaking, I'd say don't

DOI: 10.1201/9781003193081-8

FIGURE 8.1 Activating the right side of the brain brings out the creativity of engineering.

censor yourself unless, of course, there is a really good reason to do so. The point of the questions is not to annoy or offend or belittle people, but rather to help them open up some new ways of thinking. Once the door opens up for them to consider things in fresh new ways, there's no telling what unforeseen solutions might emerge to the engineering problems that they're trying to solve. Of course, in turn, we should also be open to the questions that others pose to us, as well as even asking questions of ourselves so that our own innovative thinking can be unleashed.

As a senior mechanical/biomedical engineer, I've supervised a steady stream of junior engineers, student engineers, and medical students who've worked on various research projects. I've had the habit of asking them really basic questions that might seem obvious to them and that might make me—who is supposed to be an expert in the field—seem quite dull and uninformed: How did you measure that? Why do the data points stop there? Is there any electrical interference in the lab that might have caused that weird result? Where did you find that bit of information? Did you turn the specimen upside down and try it that way? What do you think is the best way to go forward? etc. There have been so many times students have said to me, "Oh, I never thought of it that way," or something similar. As a consequence of such questions, problems were solved, new approaches were tried, serious errors were avoided, research projects were completed, and everyone learned something new.

Play Around

Another thing to do is to use play as a valid path to engineering discovery. You don't always have to limit your engineering work to linear, logical, step-by-step processes. That's not always going to get you to where you want or need to go. And you might find out quite accidentally that there are other better ways to get there more effectively and efficiently. So, once in a while at least, treat your workplace—whether it's an office, workshop, laboratory, or factory floor—as an amusement park, a sports arena, or a gymnasium, or whatever. It's a place to go to have fun with your engineering computations, drawings, experiments, etc., without any plans or pressure to get certain things done in a certain way by a certain time.

Many years ago, when I did my PhD in mechanical/biomedical engineering, my research topic was using ultrasound images to measure the stresses in total knee implants—which are made of metal and plastic—for treating osteoarthritic patients. At one point, my experimental apparatus was no more than a fish tank full of water, a bunch of metal and plastic bits and pieces, and an ultrasound probe. I was having trouble getting things to work. And I was worried. One day, my supervising professor walked into the lab. He encouraged me to do some trial-and-error tests, play around, just have fun, and let my curiosity take its natural course. And so, I tried to develop this new attitude and approach. I'm glad to report that eventually, I did get my experiments to work properly and, of course, I got my PhD as a consequence. And that wise advice from my professor stayed with me throughout my engineering career.

Play Right Brain Games

It's also helpful to engage in right brain games. This is different than just playing around without any plan or goal in your office, workshop, laboratory, or factory floor, as I discussed earlier. Rather, the games and activities I'm talking about here are intentionally designed to stimulate the right side of the brain in order to enhance your creativity. You can do some of these by yourself or in a group. There is good psychological evidence for it. So, here is a brief list of some games and activities—most of which are obvious from their names—that you can do before or during your engineering work to make you more imaginative: breathe through your left nostril; create a group story as each person adds only one sentence at a time; doodle or scribble randomly; draw a connect-the-dots picture; draw with your non-dominant hand; listen to music; play a musical instrument; play catch with a real or imaginary ball; read a fictional short story; sing a song; write down your dreams; write in mirror letters; write with your non-dominant hand.

I remember years ago when I was a teaching assistant helping a professor with an engineering design course at university. For this course, the students were always organized into small groups of 4 students each, and they sat at their own table in the large classroom. At the start of each session, the professor would ask the students to play a right brain game for, say, 10–15 minutes in order to activate their creativity. After the game was done the professor would then assign the actual engineering task that they had to complete that day. One of the right brain games I remember quite clearly was playing catch with an imaginary ball. One student would throw the imaginary ball to another student and so on. But each time one student threw the ball, they would have to speak out a nonsense word or sound that would mimic the throwing action, like blerp, boing, klomp, shoop, whoosh, and so on. Although to some people such an exercise may seem silly, it did stir up each student's visual and verbal imagination, as well as encouraging them to be free, spontaneous, and unreserved. These are helpful qualities when generating innovative engineering ideas.

Do Some Brainstorming

A very useful method for all engineers regardless of specialty is brainstorming. This is a common and effective group technique—although it can easily be used by an individual—for generating a lot of creative ideas very quickly for solving engineering problems. If you've done this with a group either in-person or online, you're already familiar with the idea. But, I'll go over some basics because brainstorming is often not done properly.

First, gather your group of engineers, but a large group needs to be split into several smaller groups so everyone can participate. Get a whiteboard, sheet of paper, laptop computer, or video/audio device to record ideas. Assign one person to record the ideas that people will speak out. Communicate to the group that all ideas are welcome no matter how bizarre since you want people to feel bold and free to speak out. State that no one is allowed to make positive or negative comments about any idea that is offered, since now is not the time for analysis. Clearly define the engineering problem or question to be addressed. Then, let the group speak out their ideas in a rapid and free-flowing way. Limit the brainstorming time to only 20–30 minutes, so people have enough time to think of ideas, but not enough time to overanalyze their own thoughts and suppress themselves. Then, you should take a break for, say, a few minutes, an hour, or even a day, and then repeat the exercise once or twice more to arrive at your final list of concepts.

However, there are some cautions to keep in mind about brainstorming. Extroverts can tend to dominate the time by speaking out more frequently, thereby intimidating the introverts. And employees may feel hesitant to speak out suggestions that could contradict their boss's ideas. So, you could also encourage everyone to write down some ideas that will be added anonymously to the final list later on. Furthermore, brainstorming can also be done fully anonymously without any group interaction by asking each individual to send their ideas to the brainstorming coordinator via paper notes, smartphone texts, emails, etc.; however, the drawback of this is that it can take away from the synergy of group sharing where one idea spontaneously inspires another idea. Eventually, the group needs to evaluate the ideas that were generated and choose the best one(s), but that is an entirely different topic that I'm not going to talk about here.

The power of brainstorming is nicely described in Guy Gugliotta's article "One-Hour Brainstorming Gave Birth to Digital Imaging" for the February 20th, 2006, edition of the *Wall Street Journal*. The article tells how Willard Boyle and George Smith had a one-hour brainstorming session in 1969, in which they conceived—and later patented—a rudimentary memory chip for storing digital information. The storage chip they invented became very attractive to astronomers who wanted better quality photos from their telescopes and the military who wanted higher resolution photos from their satellites. Over the years, the original chip was developed further and became the foundation for modern digital still and video cameras. Consequently, years later in 2006, Boyle and Smith were recognized for their achievement by being awarded the Draper Prize from the National Academy of Engineering in the US.

Welcome Mistakes

It's also important to learn from our errors. Usually, when people say something like this, they mean we should learn from our mistakes, so that we never repeat them again. Now, that's a good policy for engineers, especially when it comes to how the technologies we develop can affect public health, safety, happiness, and prosperity. But, in contrast, what I mean by welcoming mistakes is that we might discover something new and important through accidents, defects, errors, or weird results. Normally, we might be tempted to throw out specimens, prototypes, or data that don't go along with our plans, expectations, or hypotheses. I'm suggesting, instead, that we should slow down to take a good close look at those so-called mistakes. There might be a nice surprise awaiting us. But our minds must be ready to receive these surprises.

As Thomas Kuhn (1922–1996) explained in his 1962 groundbreaking book *The Structure of Scientific Revolutions*, the step-by-step process of normal everyday science usually functions within an existing overall paradigm. But, occasionally, anomalies are discovered that start to reveal flaws in the old paradigm. As more evidence accumulates, the old paradigm is forced to give way to a new paradigm, despite the objections of the critics and laggards who desperately cling to the old way of thinking. And then normal everyday science resumes within the new paradigm. Thus, he coined the term "paradigm shift." Science, therefore, progresses slowly through long periods of evolutionary steps, which are punctuated occasionally by revolutionary jumps. All because scientists—who are mentally prepared—don't simply ignore strange results. Similarly, I suggest that engineers should pay attention to mistakes that arise in their work. If an engineer analyzes and understands such mistakes more thoroughly, it could help them be more productive in the designing, building, inspecting, repairing, and/or disposing of products or services.

I refer again to Beveridge's book *The Art of Scientific Investigation*, where he states that "Although it is common knowledge that sometimes chance is a factor in the making of a discovery, the magnitude of its importance is seldom realised." (p. 43). He then gives over 30 examples of purely accidental scientific discoveries. I'll only mention a few of his examples, but add some of my own. The ancient engineer Archimedes (287–212 BC) yelled "eureka" when he accidentally discovered water weight (or volume) displacement while getting into a bathtub. Luigi Galvani accidentally detected electric current via a twitching dead frog in 1780. Michael Faraday accidentally liquefied chlorine in 1823. Louis Pasteur accidentally discovered immunization in 1880. Wilhelm Roentgen accidentally observed X-rays in 1895. Antoine-Henri Becquerel accidentally noticed radioactivity in 1896. Alexander Fleming accidentally discovered penicillin in 1928. Enrico Fermi accidentally encountered slow neutrons in 1934. Percy Spencer, an engineer, accidentally noticed the presence of microwave energy via a melting chocolate bar in 1946. Benoit Mandelbrot accidentally observed fractal patterns in electric signals in the 1960s. Robert Wilson and Arno Penzias accidentally heard unknown buzzing noises on their research antenna in 1964—which turned out to be background microwave radiation left over from the Big Bang—for which they won the Nobel Prize.

Take Breaks

Don't forget to change focus by taking breaks from time to time. Of course, it's always good to rest your mind and body from the grueling work that engineering sometimes requires. If you're tired, hungry, thirsty, and not thinking clearly this will make you less effective. You owe it to yourself and to your employer to rest, relax, and do something completely different and then get back to work with more vitality. But, there's more to it than that. There's some psychological evidence that taking breaks to do something completely different can allow your mind to keep working on ideas and solutions at a subconscious level. So, go visit your friends, play some sports, go for a walk, read a novel, watch a movie, or whatever you need to do to change your mental focus.

For instance, perhaps you've heard the well-known story about the chemist Friedrich August Kekulé (1829–1896) who was trying to figure out the chemical structure of benzene. One day while working, he decided to take a nap. He dreamed of a snake that formed a circle by biting its own tail. After he awoke, he was inspired to figure out that benzene was organized as a circle—that is, a hexagonal ring of 6 carbon atoms connected to each other—but with additional bonds sticking out so that each carbon atom could also simultaneously be attached to other atoms or rings. This idea went on to revolutionize organic chemistry.

In a similar way, consider the renowned electro-mechanical engineer Nikola Tesla (1856–1943). As he recounts in his 1919 memoir *My Inventions: The Autobiography of Nikola Tesla*, one day he went for a walk with a friend in a city park and started to recite some poetry he had memorized. Suddenly, as in a flash, came a revolutionary idea. He bent down, grabbed a stick, and started to draw a picture on the ground of a rotating wheel of electromagnetic energy. This was the basis of his invention of the brushless AC (alternating current) induction motor. This is the main way massive amounts of power have been generated and distributed to cities, industries, workplaces, and homes ever since.

Read and Watch Science Fiction

The video *Isaac Asimov: Visions of the Future* (Analog SF Magazine and Quality Video, Inc., 1992) was hosted (not surprisingly!) by the famous sci-fi and science writer Isaac Asimov (1920–1992). In the video, he asked, "how do you differentiate between science fiction and science?" In answer to his own inquiry, Asimov responded that "it's very difficult to do so because not only is the boundary fuzzy, but it changes all the time—and very rapidly sometimes—usually in the direction of science." So, because of its inherent connection to science, I'd encourage you to occasionally read and watch sci-fi. Many sci-fi books, short stories, television shows, internet shows, and movies have been, and continue to be, created. Some sci-fi writers and creators are, in fact, educated as scientists and engineers. And some of their stories are solidly rooted in the established principles of biology, chemistry, and physics. High-quality—and even sometimes low-quality—sci-fi works can inspire scientists and engineers with outside-the-box thinking and innovative technologies. It can encourage scientists and engineers to come up with new concepts, approaches, and solutions in order to advance human society.

So, for example, Jules Verne's 1865 novel *From the Earth to the Moon* was a century ahead of its time in predicting that humans would one day land on the moon via rocket ship. H.G. Wells's 1903 short story "The Land Ironclads" narrated how a small group of massive steel vehicles with automatic guns conquered a traditional army of foot soldiers defending their trench, which was published 14 years before the invention and use of tanks in World War I. Karel Capek invented the term "robot" in his 1920 play *R.U.R.* in which a factory called Rossum's Universal Robots manufactures flesh-and-blood androids equipped with artificial intelligence, which is being developed in modern times. Sir Arthur C. Clarke's 1979 novel *The Fountains of Paradise* described a tower-like space elevator used for transporting people and goods to a spacecraft docked at the tower's top, which has been researched in real-life by dozens of universities and NASA.

Similarly, the original 1960s sci-fi television series *Star Trek* and its spin-offs depicted many intriguing technologies. The "phaser" was a powerful handgun that shot laser-like energy beams to temporarily stun or totally vaporize an enemy. The "replicator" created any drink, meal, or gadget requested by the user seemingly out of thin air. The "tractor beam" was an energy tether that allowed a spaceship to capture an object in space and pull it along. The "transporter" was able to disassemble a person atom by atom in one location and then fully reassemble the person in another distant location. The "tricorder" was a portable handheld device that gave a full medical diagnosis after a quick scan of a patient. The "warp drive" could propel a spaceship at multiple times the speed of light. Some of these things may one day become reality, while others may be impossible due to the laws of physics, but can still inspire engineers to think creatively.

So, What's the "Take Home" Message of This Letter?

We engineers are usually data-driven, evidence-based, rational, and logical, so we mostly function using the left side of our brain. Step-by-step processes in our engineering work are definitely a central part of what we do. I'm not arguing against that at all. But, let's realize our need to stir up the right side of our brain to be even more productive in our work. By deliberately activating our creative forces, we will be better prepared to solve engineering problems in outside-the-box imaginative ways. And, thus, we will potentially deliver much more effective products and services to our clients, customers, and society. That's what the art of engineering is all about.

All the best,

R.Z.

Letter 9 The Business of Engineering

Dear Nick and Natasha,

Hopefully this letter finds you prospering in all things. In this communique, the focus will be on a subject that many engineers don't really "pay" enough attention to—and that's money! Now, many engineers are attracted to their profession because they think it will mainly be about using science to solve practical problems, expressing their creativity through the design process, and advancing the frontiers of technology. And, yes, it is all those things. But, to do those things, it takes money. Sometimes lots of it.

Whether we like it or not, we engineers need to recognize that at every step of the way in our journey—from studying engineering at university all the way through to retirement from engineering—money is going to be a constant companion. Consequently, to optimize our career in order to benefit ourselves, our employer, our profession, and our world, it's worth taking the time and effort to learn a few basic things about the business side of engineering.

Make no mistake, I'm absolutely not an expert on personal finance, product marketing and sales, or economic theory. But, I've done my best to learn something about the topic by taking a couple of university courses, reading books, listening to audio files, and watching videos. Perhaps just as importantly, I've had years of practical experience dealing with the business side of engineering. However, I'd also encourage you to get business advice from an educated and certified expert, in addition to reading this letter.

So, I want to highlight some common financial opportunities, challenges, and tasks that you may have to deal with at different points in your engineering career. And then, I want to tell you an allegorical, but very realistic, story about an engineer who incorporated every one of the business aspects I discuss.

Common Business Aspects of Engineering

I'm going to alphabetically list and describe frequently encountered business facets of engineering. Most can either be part of the "inflow" of money (i.e., income) or the "outflow" of money (i.e., expenses). It's all a matter of perspective (see Figure 9.1). So, for instance, salary is obviously considered "inflow" by the engineer who receives it from their employer, but the salary is considered "outflow" by the employer who pays it to the engineer. The same can be said for a product still under development vs. a product that is developed and now being sold in the marketplace, a company that hires expert consultants to advise them on technical matters vs. a company that offers consulting services to others, and so on. Let's now turn to the business facets themselves.

DOI: 10.1201/9781003193081-9

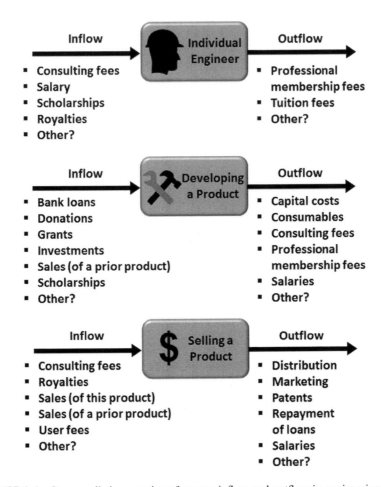

FIGURE 9.1 Some realistic scenarios of money inflow and outflow in engineering.

Bank loans are funds temporarily given by a bank to an individual, group, or organization to help pay for the equipment, facility, salaries, etc., to start an engineering business. The amount of the loan plus an additional percentage has to be paid back to the bank eventually.

BEng&Mgt stands for "bachelor of engineering and management." This is an engineering program offered by some universities that mixes a number of business courses into the curriculum. This may or may not extend the time duration of the engineering degree beyond the typical 4 years.

Budgets are detailed proposals for how an individual, group, or organization plans to obtain (i.e., income) and spend (i.e., expenses) money on engineering education, project, business, or organization.

Capital costs are the financial costs needed to purchase buildings, tools, equipment, furniture, renovations, etc., which will remain a permanent fixture of an engineering business or organization.

Conflict of interest refers to any payments, benefits, services, or relationships that an engineer receives or has received that could potentially bias the product they are developing for a client, the research they are doing at the university, the academic article they are writing for a journal, the grant application they are applying for to the government, etc. These conflicts can be real, potential, or merely perceived, and should always be admitted openly, followed by signing all appropriate forms.

Consulting fees are payments to an individual, group, or organization of engineers for providing expert technological advice and information to an organization or government that is not their regular employer.

Consumables are materials, stationery, supplies, etc., that are fully used up during the course of an engineering project. This can also include user fees, repair fees, travel costs, utility bills (e.g., electricity bill, hot water bill), etc., that are incurred during the engineering project.

Copyrights are legal rights that protect the sole ownership and potential financial rewards of an individual, group, or organization for a written or artistic work, such as an article, book, drawing, graph, photo, audio, video, or computer software. The copyright owner can take on the costs and responsibilities of marketing, selling, and distributing the resource, or they can sign a document that permanently or temporarily gives permission to another party to take on those responsibilities. In return, the creator of the work retains author/creator credit, gets discounts for purchases of the resource, and receives royalty payments from the other party.

Donations are voluntary and permanent contributions of funds, goods, or services from an individual, group, or organization to partially or fully support an engineering project, business, or organization. These do not have to be paid back to the giver. Donations usually have few, if any, specific expectations or restrictions on the work that will be accomplished compared to grants (see later).

Expenses are the costs incurred to pay for the consumables, marketing, salaries, taxes, etc., of a project, business, or organization. This is the money "outflow" part of an engineering budget.

Grants are funds, goods, or services obtained for an engineering project by an individual, group, or organization that formally applies to the grant program offered by an organization or government. In turn, grant programs are made possible through donations of private individuals, an organization's own charitable giving plan that is built into their budget, or government public taxes. These do not have to be paid back to the granting agency since they are permanent contributions. Grants often have very specific expectations and restrictions on the work that will be accomplished compared to donations (see earlier discussion on *Donations*).

Income is the money received by an individual, group, or organization through product sales, donations, grants, royalties, scholarships, etc. This is the money "inflow" part of an engineering budget.

Investments are voluntary, but temporary, loans of funds, goods, or services from an individual, group, or organization to partially or fully support an engineering project, business, or organization. The amount of the investment plus an additional percentage have to be paid back eventually to the investor.

MBA stands for "master of business administration." This is a university degree that some engineers decide to get to help them better understand the business aspects of engineering, to start their own engineering firm, or to prepare them to move into a management role with their employer.

NDA stands for "non-disclosure agreement." This is a morally binding, but not necessarily legally binding, contract that an engineer, business, or organization signs. All signatories promise they will not privately or publicly leak any information to non-participants about the engineering research or product development being done until they agree otherwise. Failure to comply could lead to lawsuits, but only if the offended party has the money and means to start legal action against the party that broke the agreement.

Patents are legal rights that protect the sole ownership and potential financial rewards of an individual, group, or organization for an engineering concept, physical product, computer software, or technique they invented. Patents are obtained either for the appearance and/or function of the invention. There may be financial costs associated with applying for and maintaining a patent.

Professional membership fees are the costs a person pays to maintain their membership status in a professional engineering society. University engineering professors and senior engineers in industry or government are usually required to have such memberships by their employer. In some cases, the engineer's employer will pay these fees directly or reimburse the engineer, but some employers don't do this.

Royalties are one-time or ongoing payments from an individual, group, or organization to another individual, group, or organization for permission to use a legally-owned patent, copyright, or trademark. Some examples can illustrate how this works. In the case of patents, a company can pay a fee to a patent owner for permission to make, market, and sell a product based on the patent owner's idea. In the case of copyrights, a book publisher can pay the author of a book a fixed percentage of the book's sales for permission to produce, market, and sell the book. In the case of trademarks, an organization can pay a fee to a trademark owner for permission to use a name, word, phrase, or symbol that legally belongs to the trademark owner.

Salaries, as well as benefits packages and retirement plans, are the monetary compensations that engineers receive as employees of an organization or government, or that engineering students receive from their supervising professors while they are completing a master's or doctoral degree.

Scholarships are gifts of funds to support an engineering student—whether or not they are also employed—who formally applies to the scholarship program offered by a university, organization, or government. In turn, scholarship programs are made possible through donations of private individuals, an organization's own charitable giving plan that is built into their budget, or government public taxes. These do not have to be paid back to the scholarship program, since they are permanent gifts.

Trademarks are legal rights that protect the sole ownership and potential financial rewards of an individual, group, or organization for a name, word, phrase, or symbol that is used to identify the individual, group, or organization and, hence, its products and services. There may be financial costs associated with applying for and maintaining a trademark.

Tuition fees are costs an engineering student pays to the university while completing an engineering degree. Students studying in another country (i.e., international students) usually have much higher tuition fees than students studying in their own country (i.e., domestic students). These fees help partially cover the costs of professors' salaries and physical upkeep of the university campus. In some cases, an employer will support one of their engineers who's still working to study for a master's or doctorate in engineering and will usually pay these fees directly or reimburse the engineer when they complete their degree.

The Story of the Engineer Who Took Care of Business

Once upon a time, there was a young kid who always liked taking apart toys and then putting them back together again. In high school, her favorite subjects were computers, economics, mathematics, and science; she always got excellent grades in her courses. When it came time for university, she knew she wanted to be an engineer. So, she applied to some universities and, gladly, she was accepted by one that even gave her an entrance scholarship because of her excellent high school grades. Of course, she knew she would use some of that scholarship money to pay for her university tuition fees. She enrolled in the BEng&Mgt program because she wanted to start her own engineering business one day. She excelled at her studies over the next few years. And she finally completed her degree.

After that, she put in a lot of time and effort to create a really nice resumé. She started applying for jobs and going to interviews. She was soon hired by a small engineering company. Her job was designing widgets, which she did to the best of her ability. Her boss saw what a good worker she was and that she previously took some business management courses at university. So, the boss promoted her to lead a team to design and make a new kind of widget. She was responsible to make sure the team stayed within a budget, so the expenses of making the widget would be less than the income from selling the widget.

But, to prevent some other company from stealing the new widget design, the company applied for and got a patent. Now, they could be the only ones in all the land who were allowed to make, market, and sell the widget. And the company also made sure the engineer and her team signed an NDA, so that they wouldn't accidentally tell their friends or strangers about the big secret of how they really made their widget.

After many enjoyable years working for the company, the engineer decided she wanted to go back to school to get an MBA. At school, she made a few good friends who were business students. They all came up with an idea for a new type of gizmo that would solve a practical societal problem. And then they decided to start their own small engineering company to produce the new gizmo they invented. They knew this would be a big effort, and they would need money to get their business started. So, they successfully got a bank loan, a few private donations and investments from family and friends, and a government grant for small start-up businesses. This helped them pay for the capital costs of a small building and some important equipment for making the gizmo.

Next, they hired new employees to whom they paid decent salaries. They also made sure to include in their budget the consumable costs for buying the raw materials from which each gizmo was made. Then, the engineer and her team got a patent for their new gizmo idea and legally registered the gizmo's trademark name and symbol, so they would be the only ones who could legally make, market, and sell it. Before they could start their business, the government told the engineer that she needed to be a recognized member of a professional engineering society. The society made sure that she and all other engineers across the land made products that were good quality and safe for their customers to use. She passed the required exam and started paying her membership dues.

She was ready to start her business. And what a business it was. Every year they sold more and more gizmos. The business was a booming success! But, despite all the skyrocketing gizmo sales, they wanted to scientifically prove their gizmo was better than any other products that claimed to do a similar thing. So, they got a government grant to fund a gizmo research project. Happily, the results showed that their gizmo was better than its commercial competitors. They even published a scientific article on their gizmo in a widely read journal. However, the journal required the engineer to sign a conflict of interest form. This form allowed the journal to let its readers know there was a potential bias in favor of the gizmo in the article because the engineer's business did the research project on their own gizmo.

After many happy years running her own successful engineering business, the engineer finally retired. She wrote a book about her life and work. She signed a contract with a book publisher which transferred the copyright to them, but she received many book royalty payments from the publisher because the book sold so well. The book inspired so many new up-and-coming engineers, that she was invited by them to be a paid freelance consultant to help them kick-start their own engineering firms. The engineer felt a deep sense of satisfaction knowing that she was leaving a legacy that would help the next generation of engineers continue to build civilization. And she lived happily ever after!

So, What's the "Take Home" Message of This Letter?

The aim of this letter was to point out that an engineer should understand the real impact that money has on their career. This doesn't mean that we engineers have to be experts in banking, economics, or finance. This doesn't even mean that we engineers should start our own engineering businesses. But, what it does mean, is that we engineers can have much more successful and satisfying careers if we learn how to leverage the business aspect of engineering to benefit ourselves, our employers, and society at large. So, let me encourage you to get a hold of some good books, watch some useful videos, take a university course or two, and get some advice from a properly educated and certified business expert. And, if necessary, even get an MBA or similar degree, so that you can be fully prepared to meet the challenges of tomorrow!

Kind regards,

R.Z.

Letter 10 The Ethics of Engineering

Dear Natasha and Nick,

Let me start this letter by asking you a question. Did you ever wonder why it sometimes may seem tough to optimize competence and character in our engineering careers? This is humorously illustrated in the television comedy series *The Big Bang Theory* by the devious engineer Howard Wolowitz. In one episode, he tries to impress a friend whom he sneaks into a government-run Mars Rover control room, but then accidentally gets the vehicle stuck in a ditch. Instead of admitting his mistake, he recruits his colleagues to help him in a comical series of deceptions.

But, more seriously, competence and character are both especially important for engineers, since our work can positively or negatively affect the health, safety, happiness, and prosperity of society at large. Unfortunately, just like everyone else, we engineers can think, say, and do things we regret later. Ideally, we'll have the chance sooner or later to take responsibility, learn some lessons, make amends, and never repeat those errors again. And that's what this letter is about, namely, the ethics of engineering.

Now, the word "ethics" could be reasonably defined as the study of what people think is either morally acceptable or unacceptable human behavior. Of course, this topic is a murky one, since how this works out in people's daily lives has been debated for centuries by philosophers and theologians. Even so, when it comes to engineering, there are concepts and tools that can help us do our work ethically.

Before I really get into the subject, let me state a few caveats. First off, I readily admit that I haven't personally figured this out, or even lived it out, fully and perfectly, but I'm still learning. Also, each region, country, and professional engineering society (PES) that legally certifies a professional engineer (PE) has slightly different philosophies, policies, and laws to make sure engineers do their work ethically. And, there are comprehensive resources specifically on engineering ethics that you should get your hands on. With that said, I hope what I write below inspires you to dig into this topic even more.

The Wheel of Engineering Ethics

Let's start by discussing some ideas that should be kept in mind by every engineer who wants to work ethically in order to benefit their customer, client, employer, profession, and society at large. Since metaphors and visuals have always helped me grasp concepts that might otherwise be fuzzy, I'd like you to picture what I like to call the "Wheel of Engineering Ethics" (see Figure 10.1). Each of its 4 spokes symbolizes an equally important aspect of engineering ethics. So, ideally, all 4 spokes will be in good shape so the wheel can turn and go forward at full speed.

DOI: 10.1201/9781003193081-10

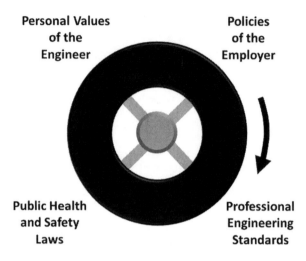

FIGURE 10.1   The wheel of engineering ethics.

But, these 4 spokes also function like a system of checks and balances. Even if just one spoke is broken (i.e., something's unethical based on just this criterion), the wheel will warp and have difficulty turning and moving forward (i.e., the unethical project will have difficulty getting done).

The first spoke represents *personal values of the engineer*. In our personal and professional lives, most of us recognize that our attitudes lead to our actions, our beliefs determine our behaviors, and our principles inform our plans. We don't always painstakingly take the time and effort to analyze and reanalyze every little daily decision we make to uncover the underlying core reason for it. If we did, we'd never get anything done. We may not always be consciously aware of why we think, say, or do certain things, but there are core values that undergird it all. But, how many of us have actually taken the time to think about, write down, thoroughly analyze, and even adjust our core values as engineers? Our values could be influenced by certain philosophical or religious ideas or just by our own life experiences. Whatever the case may be, it can be a useful exercise for all of us to do at least once in a while, since it can help us say "yes" or "no" to the various engineering questions we need to answer with more ease and conviction: Should I take that job? Will I accept that project? Should I report those results? Am I truly qualified to design, build, inspect, repair, or dispose of this product? Shall I express my concerns about the health and safety of this technology? etc. After all, even if the whole world says it's okay to do something, but we think it's unethical and just don't feel right about it, then we have an extremely tough question to answer—will we follow the world or will we follow our conscience?

The second spoke refers to *policies of the employer*. The 3 primary places of employment for engineers are the university, industry, and government. At some point after hiring an engineer, a responsible employer will provide the engineer with information—such as a website, computer files, or some booklets—about the employer's policies regarding both the employer's and the engineer's duties. And this includes the employer's and engineer's duties to behave ethically in the workplace towards others, but it also involves employee health and safety and the engineer's technical work itself. If this information isn't given right away, the engineer should approach the employer—say, the human resources (HR) officer or department—to get the information, so they know about the employer's policies. Of course, if the employer is a small business, they might not have an HR officer or department or even a comprehensive information package that includes a code of conduct dealing with ethics. In this case, the engineer can request the employer to provide a simple document on ethics, ask the employer if they'd be willing to adopt the code of conduct from a PES, or some other idea. The point is that the engineer needs to get something in writing to clarify the employer's and their own duties; don't just rely on a verbal agreement. On the one hand, if the engineer fails to abide by the employer's policies, of course, they could be demoted in position or dismissed from work altogether. On the other hand, many employers are so driven by financial profits, impending deadlines, or commercial rivalries that an engineer may feel forced to engage in behaviors or projects that seriously breach their own personal values, the employer's own policies, professional engineering standards, and/or public health and safety laws. If appealing to the HR officer or department, the team or department leader, the president of the company, etc., doesn't resolve these issues, then the engineer has other options—just get back to work, quit the job, or become a "whistleblower." A whistleblower is someone who bypasses their employer and reports the problem directly to their PES, the public media, the legal system, and the government to discourage the employer from continuing the unethical behavior.

The third spoke signifies *professional engineering standards*. If you've been formally certified as a PE by a PES in your region or country, then you're an ambassador of the PES and, thus, are obligated to understand and abide by its standards, such as its code of conduct. A code of conduct typically emphasizes an engineer's responsibility to do their work using concepts like accountability, confidentiality, compassion, competency, fairness, honesty, legality, loyalty, objectivity, respect, and trustworthiness. If you deliberately or ignorantly fail to comply with the PES code of conduct, you could be temporarily or permanently stripped of your legal PE certification. Now, even if you're not a PE, but you are a working engineer, it's good to become familiar with a code of conduct to help you decide whether something is ethical or not, as well as to understand the views of your peers in the engineering profession. Moreover, technical competency is also a vital part of being an ethical engineer. So, you should be aware of, understand, and incorporate, whenever possible, the widely recognized international engineering standards that describe the technical specifications required for material quality, experimental testing, product performance, health and safety levels, and so on. The most commonly used standards are the American Society for Testing and Materials (ASTM) and the International Organization for Standardization (ISO), although there may be some local standards that are frequently used in a particular country. If the ASTM and ISO don't have documents that address your need, then consider looking to manufacturer guidelines, journal articles, textbooks, handbooks, manuals, and codes to provide a precedent or guideline for formulas, procedures, standards, and so forth.

The fourth spoke means *public health and safety laws*. It's often said that technology can either help or hinder the health and safety of human society and the ecology. Although that's easy to see in many cases, often it's not a binary choice and it's more complicated than that. So, for instance, maybe technology is greatly beneficial in one way, but it also has some serious negative side effects. What should be done in that case? Should the technology be permitted as it is, banned outright, or only used in certain situations? Or, perhaps, is it worth the time, effort, cost, and resources to improve or replace that technology? How and when will these issues be resolved? And who will decide? etc. These and other philosophical questions are faced by engineers and society all the time. And the answers that are given in response to these questions can have serious consequences in the here and now, as well as to the future survival of human civilization and our planet. Consequently, most developing societies have local or regional laws (e.g., building codes for structural integrity, electrical wiring, plumbing systems, waste disposal, etc.) and nationwide laws (e.g., government regulatory agencies that approve or reject applications for patenting, marketing, and selling of technologies) that address public health and safety. Obviously, these laws need to be balanced with other factors like financial prosperity, personal choice, public opinion, cultural tradition, and so forth. The ethical engineer and engineering organization need to know and abide by these laws because they affect how they do their work, otherwise, they could face boycotts, closures, dismissals, fines, lawsuits, or prison.

True Tales of Engineering Gone Wrong

The short vignettes below are meant to illustrate how an engineer's poor ethics can have bad and sad consequences. Through negligence, incompetence, or malevolence, an engineer's shoddy work, an engineer's conscious decision to develop or misuse technologies, or the misuse of technologies by others can cause illness, injury, discomfort, death, or destruction. In many cases, the engineer, their employer, and other participants can potentially face bad reputations, stripped credentials, financial penalties, lawsuits, or prison. It's disturbing that many similar cases have happened in history. Each event, whether quiet or dramatic, is a poignant reminder to engineers about their moral obligations to themselves, their employers, their profession, and society at large.

The first story is about the *explosion of a famous spaceship*. I personally remember watching this tragic event on the television news. What happened? As reported in physicist Richard Feynman's book *The Pleasure of Finding Things Out*, on January 28, 1986, a short time after NASA launched the Space Shuttle *Challenger*, it exploded in mid-air killing all 7 passengers. The cause was later identified as the failure of important O-rings (i.e., gaskets) on the shuttle's solid rocket booster. Here's how this was discovered. Following the calamity, an official Commission was set up to determine the cause of the disaster and to recommend changes to prevent future ones. The only scientist assigned to the Commission was Richard Feynman himself, who was a Nobel Prize-winning physicist. After a thorough investigation, which also involved interviewing the engineers who worked on the project, Feynman wrote a summary that was almost excluded from the Commission's final official report. He noted several important things. First, prior successful Space Shuttle flights showed O-rings with erosion depths of one-third the radius, so it was assumed that the deeper erosion to one-half the radius that was known to cause O-ring failure would not occur for the upcoming *Challenger* flight. Second, pre-launch mathematical modeling of O-ring erosion was done, but only the best-fit-curve running through the center of the scattered experimental data points was considered when making decisions about safety; however, experimental data points that were far above or below the best-fit-curve, as realistic extreme cases, were ignored. Third, the mathematical modeling was based on a material that was similar to, but not exactly the same as, the material from which the O-rings were made. It's disturbing to know that engineers had warned the mission's management team that the O-rings might fail due to the low temperatures on the shuttle's outside, yet the launch went ahead with terrible consequences. At a live press conference later organized by the Commission, Feynman clearly illustrated the problem when he dipped one of the shuttle's O-rings into a cup of ice water, pulled it out, and then easily broke it apart. And so, were the engineers unethical for not being more forceful with management, was management unethical for going ahead with the mission regardless of safety concerns, or both?

The next tale concerns the *orthopedic implant industry*. I'm quite familiar with this topic because of my many years working as a hospital-based and university-affiliated mechanical/biomedical engineer. Orthopedic implants are used to replace bad joints (e.g., prostheses for shoulders, elbows, hips, knees) or repair broken bones (e.g., plates, rods, screws, cables). They are made from metals, plastics, and/or fiber-reinforced composite materials. The companies that design, make, and sell implants often do some in-house stress testing to ensure they work as expected, they obtain official approval from a government regulatory agency, and then they sell the implants to doctors, clinics, and hospitals for use in patients. But, a major problem exists, namely, that the proprietary nature of design and testing data, as well as the rivalry between companies, prevents this information from being published openly in peer-reviewed engineering or medical journals. Sadly, without much publicly available scientific evidence available to them, surgeons sometimes choose, buy, and insert unproven implants into patients. And so, mechanical and biomedical engineers like my colleagues and me—who have no formal connection to these companies whatsoever—have to rush to perform computational analyses, experimental stress tests, and patient trials in our labs on these unproven implants and publish our findings in academic journals for all to read. We want our articles to finally provide comprehensive engineering data about the quality and performance of those previously unproven implants. Only then can surgeons make good clinical decisions for their patients. As a consequence of this somewhat backward system (i.e., implants are approved and used before they're fully proven), many implants have caused pain, discomfort, infection, wear debris, and bone fracture in patients. Not surprisingly, there have been many highly publicized lawsuits against implant companies resulting in millions of dollars in fines and banning of particular implants from the marketplace. So, who's at fault here? Is it the scientists and engineers who design the implants but don't convince their employers to allow them to publish the results openly, the government regulatory agencies who approve the implants before they're totally validated, the implant companies who sell the implants that haven't been fully proven yet, the surgeons who insert the implants in patients without comprehensive evidence, or all of them?

The final example is about an *engineering student who cheated*. Many years ago, I was a teaching assistant helping a professor to run a university engineering course that was composed purely of lab exercises. The students worked in small groups, yet each student was asked to submit their own individual written lab report for grading. When I started to grade the lab reports, it was very obvious to me that one student group didn't write individual lab reports, but they copied from each other often word for word. They cheated. So, I gave each of those particular students the lowest possible passing grade of 50%. I told this to the course professor, who said I was too lenient; the professor would have divided the grade by the number of students in the group or just given them each a zero. Then one—and only one—of the cheating students came to me to complain about the 50% grade. After some discussion, the student admitted he cheated. But, what was quite disturbing, was that he didn't think cheating was a serious issue. He didn't express any regret. His justification was that the university really wouldn't care about him after he graduated anyway, so engineers like him and me needed to stick together. I was mildly shocked that he tried to recruit me for his unethical attitudes and actions. I kindly told him he was wrong and that I was perhaps too generous in giving him a 50% grade since the course professor wanted to give him an even lower grade. He went away unconvinced and unhappy. What eventually became of him in later years, I don't know. But, I can only hope that my little conversation with him somehow changed his mind before he became an engineer in the workforce who could affect the health, safety, happiness, and prosperity of society at large. So, who's to blame in cases like this? Is it the engineering educational programs that pressurize engineering students with overly busy schedules and yet expects them to perform well, the professors and teaching assistants who don't communicate persuasively enough about engineering ethics to their students, the individual engineering students who don't take personal responsibility for their own actions, or all of them?

So, What's the "Take Home" Message of This Letter?

To bring balance to technical competence, which is the major focus of engineering education and training, it's necessary to add the element of moral character. But, since the personal values of an individual engineer cannot be solely relied upon by society, it's vital for an engineer to also consider and comply with the policies of the employer, professional engineering standards, and public health and safety laws. When these 4 spokes are in place on the "Wheel of Engineering Ethics," then the profession of engineering can more confidently be relied upon to deliver products and services for everyone's benefit.

All for now,

R.Z.

Letter 11 The Personal Traits of an Engineer

Dear Nick and Natasha,

I hope everything is going well for you these days. At this time, I'd like to share some ideas about the personal traits of a typical engineer. I don't like stereotypes, but sometimes there's a grain of truth to them. Perhaps you think this is an odd topic. But, let me explain what I don't mean and what I do mean. What I don't mean is the engineer's moral character, that is, their moral traits like bravery, compassion, honesty, humility, kindness, self-sacrifice, and so on. These things are fine in themselves, but they don't necessarily make someone perform better or worse as an engineer from a scientific or technical point of view.

Rather, what I do want to talk about is the engineer's technical competence, that is, those personal traits that often do have a very direct influence on the engineer's ability to perform their job at a high level of quality. So, what I'm talking about here is competence, not character. Although I haven't done a psychological or behavioral study, I'm going to suggest 12 personal traits that, from my personal experience and observations over the years, are especially evident in high-performance engineers. You can surely think of other attributes too, but the ones I'll discuss are more than enough to make my point (see Figure 11.1).

Curiosity

The ancient Greek philosopher Socrates (470–399 BC) was known for his skillful use of asking questions to get his listeners to think more deeply about their own assumptions and ideas. This is now known as the Socratic Method of teaching and learning. In a similar way, today's engineers should have a good amount of personal inquisitiveness about all sorts of things, which often shows up in the questions they pose to themselves and others. How does that gadget work? Can I take it apart and see how it functions? Is there a way to fix this device so it works much more efficiently? What gaps exist in the academic literature of my field that need to be filled? Has anybody comprehensively studied this topic before? Is there a brand new way of thinking about this situation? Maybe I can invent a device that solves that problem? What is the state-of-the-art technology for measuring this and manufacturing that? How can I learn to do this too? Is there a fundamental assumption we're making that needs to be rethought? etc. It's this kind of genuine interest that drives many engineers to research unexplored topics, develop new products for clients and customers, and gain new knowledge and skill so they can be more effective. An apathetic engineer will be uninterested in these things and, eventually, it will manifest itself in shoddy workmanship. And mediocre engineering work can be especially dangerous for public health and safety.

DOI: 10.1201/9781003193081-11

FIGURE 11.1   Engineers need to have strong personal traits to succeed.

Creativity

An engineer should have an active imagination, since many engineering problems require new ways of thinking that lead to innovative solutions. Obviously, every individual is unique, so some engineers will naturally be more creative than others. But, why is that? There may be internal and external factors that can help or hinder an engineer's creativity. For example, internal factors involve the engineer's personal mindset, such as embracing or fearing failure, willingness or unwillingness to try new things, ability or inability to mentally visualize images, and so on. Similarly, external factors involve the engineer's immediate surroundings, such as a boss or workplace that is freeing or controlling, a particular specialty of engineering that is innovative or decaying, a culture that has few or many expectations, and so forth. For engineers who are willing and able, there are techniques that can help develop one's mental capacity for creativity and openness to new ideas. For instance, these well-established techniques include group brainstorming, playing psychologically designed right brain games, changing mental focus to allow the subconscious mind to keep working on the problem, learning new lessons and making accidental discoveries from apparent errors or mistakes, reading and watching science fiction to get inspired by unusual futuristic ideas and technologies, etc. In contrast, an engineer who is only willing to consider one possible type of solution or only one technique to arrive at a solution may actually be compromising their work and, hence, the products and services they eventually provide to their clients, their customers, and society at large.

Skepticism

The Royal Society was officially founded in Britain in 1662 and is the oldest surviving scientific organization in the world. Its official motto is "Nullius In Verba," which is a Latin phrase that means "Take nobody's word for it." It expresses the idea that it's important to find out scientific truth based on actual evidence. For many prior centuries in the western world, the views of the Greek philosopher Aristotle (384–322 BC), among others, were authoritative. Yet, his methods of inquiry—although he did promote theorizing plus experimentation as a path to discovery—were not always in-line with what we would consider good science today. Even today, unfortunately, we can tend to unthinkingly accept a prevailing paradigm or word of an expert without looking into things for ourselves. I'm not suggesting that all paradigms or experts are bad or wrong, but I am recommending we do our best to always ask, "What's the evidence for that?" If we do, our engineering work will surely improve.

Objectivity

Physicist Richard Feynman's book *The Pleasure of Finding Things Out* tells the story of the 1986 midflight explosion of the Space Shuttle *Challenger* caused by faulty O-rings (i.e., gaskets). Despite the warnings of engineers about the potential of project failure, NASA managers decided to go ahead with the launch, in part, because of timeline and funding pressures. Similarly, an engineer can also be faced with various pressures, like customer requests, project budgets, delivery deadlines, employer expectations, production quotas, and so forth. It's extremely tempting, then, to take shortcuts that could compromise the quality of our work to accommodate all these other factors. So, for instance, we might use a cheaper, but weaker, material in order to satisfy the budget. Or, we might do fewer laboratory experiments that are meant to ensure the product's quality so we can meet a delivery deadline. Or, perhaps, we might ignore certain data points on the graph since they don't agree with the original hypothesis we or others were expecting. The problem, obviously, is that we're letting these other issues negatively influence our ability to make impartial and unbiased decisions based purely on scientific facts and technical reasons. And so, for us to really do high-quality engineering work, it's vital that we maintain our evidence-based objectivity.

Orderliness

When I did my engineering studies, I had a professor who could reach back with one hand to his bookshelf without even looking at it, grab the exact book he needed, and then open it up quickly to the page he wanted to look at. He knew where everything was in his office. I remember seeing this firsthand. There's no doubt that this approach served him well, although perhaps few engineers are like that. Even though this kind of personal organization is one aspect of the orderliness I'm talking about, there's another equally important part. And that's a systematic approach to our engineering tasks. It means thoroughly reviewing the existing patents or the latest academic literature to ensure our product or project will be unique. It means systematically changing the input parameters in our research study to see the effects on the output variables. It means precisely following the step-by-step instructions of a technical manual for a new piece of equipment we've purchased for our laboratory or machine shop. Whatever the exact applications or circumstances, we engineers should attempt to bring to our work an orderly mind and an orderly approach.

Focus

One day, the famous ancient Greek engineer Archimedes (287–212 BC) was reportedly absorbed in thought over some mathematical diagrams he'd drawn in the sand on the floor of his study. A soldier from an invading Roman army burst into the study and accidentally stepped on the diagrams. Archimedes replied, "Fellow, do not disturb my circles!" and begged the soldier to let him finish his calculations. But the soldier—insulted and enraged at being told what to do—swiftly slew Archimedes. Although this tragic tale shouldn't be promoted as prudent behavior for engineers, it does hint at the kind of focus that is required. I think the lesson for us is that an engineer should be able to concentrate for long periods of time on a single problem without getting distracted by other tasks or people. Only in this way can we understand the complete nature of a problem in all its aspects and, then, hopefully, arrive at an appropriate solution. Of course, sometimes it's not just about having or developing a capacity for focus, but it may also require us to deliberately create an environment free of external distractions.

Persistence

Engineers can benefit from being a little bit stubborn. It means we don't give up when faced with a problem. It means we find ways to go around unexpected obstacles. It means we plod on to find a solution when others move on to something else. It means we work hard when others are bored or tired or distracted. In my own modest way, I've tried to do this, although not perfectly. One example is my master's thesis in mechanical engineering. My supervising professor and I published a research article in a peer-reviewed academic journal based on the core results of my thesis only a few years after I completed the degree. But, I always felt there was another potential article based on some of my other unused data; this bothered me for years afterward. And so, when I had some extra time and energy about 13 years later, I looked through my master's thesis carefully and decided to write another journal article and submit it to a journal. The paper was rejected outright by the journal with major criticisms from the reviewers, but I didn't give up. I obtained permission from the journal editor to rewrite the paper thoroughly based on the comments of the reviewers. The paper this time wasn't rejected outright, although it still received more heavy criticisms, but I still didn't give up. I then made another thorough revision of the paper and resubmitted it to the same journal. This time it was accepted outright with very encouraging comments from one of the reviewers and then published soon afterward. This was a full 15 years after the publication of the first paper from my master's thesis.

Hands-on Practicality

It's wonderful when engineers use computer programs and mathematical formulas to solve problems. But, if this is all they do, then they are divorced from the practical realities of the real-world problems they're trying to solve. And so, it's essential for engineers to be interested and skilled in using their own two hands to use tools, instruments, and machines to make things, fix things, measure things, etc. For example, early in my career when I was a junior engineer in charge of a bio-medical engineering lab, I learned how to find, purchase, install, use, and/or repair a variety of machine shop equipment and research instruments. Then, later when I was a senior engineer and professor hiring and recruiting engineering staff and students, I looked for people who also had hands-on skills because I knew it would help them in their duties.

The scientist Robert Hooke (1635–1703) is also a relevant example. He was a key figure in founding the Royal Society, which I mentioned earlier, but he was also its first Curator of Experiments. In this role, Hooke fabricated all sorts of new scientific instruments and then demonstrated how they worked to Royal Society members at their weekly meetings. Hooke invented or substantially improved the air pump for creating a vacuum, anemometer, balance spring for watches, the compound microscope, the dial barometer, hygrometer, iris diaphragm later used for cameras, and reflecting telescope. But, of course, he is mainly familiar to engineers because of Hooke's Law of elasticity, whereby he discovered by experiment that a spring will stretch by an amount that is linearly proportional to the amount of weight hanging from its end. This law is expressed today in a generic form applicable to an object of any shape or material, as $\sigma = E\varepsilon$, where σ is the stress (i.e., force per unit area), E is the modulus of elasticity (i.e., constant of proportionality), and ε is the strain (i.e., change in length per original length).

Self-Motivation

A good engineer will be a self-starter who doesn't need to be micromanaged or pressurized to do their work accurately, precisely, thoroughly, and so on. They want to do it because they have an internal drive to do it, and they see its value to their clients, customers, and society at large. But, an engineer who doesn't have this personal quality may eventually start to complain, dislike their work, fall behind in their progress, and produce low-quality products or results. I realize there can be other reasons why an engineer displays negativity and lack of performance, such as a bad boss or personal problems. Although these issues need to be resolved, they're not the focus of this letter. Instead, I'm talking about the basic internal motivation an engineer does or doesn't have.

In this regard, the Renaissance genius Leonardo Da Vinci (1452–1519) is an example for us. He was not just a master artist who created the *Mona Lisa, The Last*

Supper, and the *Vitruvian Man*, but he was also a civil, mechanical, and military engineer whose ideas were often ahead of their time. He filled his notebooks with original drawings and blueprints for axles, bridges, canal diggers, catapults, city layout plans, crossbows, fortresses, gears, hand grenades, hoists, human-powered flying machines, levers, machine-type guns, mirror grinders, missiles, mortars, needle grinders, odometers, pulleys, ratchets, screw jacks, siphons, springs, steam-powered cannons, submarines, tanks, water wheels, etc. Sadly, he never published or even built many of his ideas, in part, due to his other responsibilities, the lack of a financial backer at times, and the technological limitations of his era. But, his self-motivation for engineering is self-evident.

Collegiality

Some engineers in centuries past—like Archimedes and Leonardo Da Vinci—often toiled away by themselves to invent many of the devices, machines, and systems that built their societies. But in today's world and the foreseeable future, engineers need to partner with others in order to achieve common goals. Many engineering marvels in the modern era are so complex—like airplanes, high-speed trains, nuclear power plants, and the International Space Station—that their design, construction, and maintenance require the efforts of teams of engineers, often from multiple disciplines. And for those teams to function optimally, it's best if there is an atmosphere of mutual respect, open communication, and resource sharing. It requires a certain amount of openness to others and some social IQ on everyone's part. However, this doesn't necessarily mean that everyone will be fond of everyone else or that they will become close personal friends after the project is completed, but it does mean that a degree of civility must prevail on the team.

As a mechanical/biomedical engineer, I've had my fair share of working with many teams of electrical engineers, machinists, materials engineers, mechanical engineers, orthopedic surgeons, prosthetists, statisticians, and/or technicians to design and fabricate various medical implants and other devices. Some of these collaborators were at the same institution and/or department where I worked, while others were at entirely different institutions. Everybody brought their own particular expertise to the table without which the project would not be possible. And we all learned something new—whether it's new information or a new skill—that we could then employ later in our own work. It seems, after all, that there really is a nugget of truth to the old saying that, "Many hands make light work." Just as importantly, working with others on collaborative projects was also a beneficial move for me, since it built my professional network which I could later call upon if I needed something or, of course, vice versa. So, another lesson here is that your professional network can provide you with all sorts of assistance, opportunities, and resources in the future.

Shrewdness

Engineers need to know more than just the scientific or technological side of their profession. They need to know how to cleverly and skillfully engage with people and systems. So, for instance, it's useful to understand the human psychology of why other people make or don't make certain decisions. It's beneficial to get to personally know the right people in positions of power and influence. And it's helpful to be aware of all the ins and outs of how a workplace functions and how a certain profession as a whole operates. My point is that engineers should understand how to diplomatically, ethically, and legally leverage the various opportunities and even obstacles that come their way to their own advantage. This can help them to be more satisfied and productive in their engineering work which, in turn, will also benefit their team, employer, profession, and society at large.

As a case in point, I know about a number of engineering professors who worked at various universities. However, they were somewhat unsatisfied with their salary, benefits, resources, lab space, and so on. So, they applied for jobs to work at other universities or industries, which could better meet their career goals. Some of these professors were genuinely willing to switch to another university or industry, but only if they were offered a new job with a better deal than they currently had. However, some of these professors had no intention of leaving their current job; they only wanted to use a new job offer as a negotiating tactic to get a better deal from their current workplace. There is nothing improper about these actions since they are within the rules and they happen all the time. Even so, you don't want to do this sort of thing too often in your career, since you might get a bad reputation as a "job jumper" or as someone who plays "mind games."

Joyfulness

An engineer does not have to be a cheerful, optimistic, or pleasant person who's always smiling, chuckling, or shaking hands. Of course, there are some engineers, just like other people, who have this kind of natural emotional disposition. Rather, what I mean by joyfulness is the sheer pleasure we get from engaging in all manner of engineering tasks, as well as the deep satisfaction of knowing that we are contributing something worthwhile to humanity. The tasks may be small and seem insignificant, or they may be grand and seem crucial. In either case, the joyful engineer greatly delights in all of it.

The distinguished electro-mechanical engineer Nikola Tesla (1856–1943) wrote the following in his book *My Inventions: The Autobiography of Nikola Tesla* (Hart Brothers, 1919/1982):

> [An inventor] finds ample compensation in the pleasing exercises of his powers and in the knowledge of being one of that exceptionally privileged class without whom the race would have long ago perished in the bitter struggle against pitiless elements. Speaking for myself, I have already had more than my full measure of this exquisite enjoyment, so much that for many years my life was little short of continuous rapture. (p. 27)

In contrast, if an engineer really isn't passionate about science and technology—perhaps they studied engineering to please their families or just to make money or to satisfy their ego—then it's not likely that they're going to advance upwards in their careers, perform extremely well at their jobs, or even stay in this career for a long time. And so, engineers of the world, I implore you, take pleasure in the work of your mind and hands!

So, What's the "Take Home" Message of This Letter?

My point here is simply that a productive and fulfilling career as an engineer does require certain personal characteristics. Not every single engineer is going to have all of the ones I discussed in exactly the same measure, or express them in exactly the same way. Yet, I'm convinced that most engineers have many of these personal traits in common with each other. And if you feel you're lacking in a few attributes that might be important for your particular career plans, then I'd encourage you to develop them as much as possible. I'd recommend looking into books, videos, websites, mentors, seminars, and other similar resources on personal development and empowerment that have been created by true experts in this area.

Peace and prosperity,

R.Z.

Part IV

Workforce

Letter 12 Where Do Engineers Work? University, Industry, Government

Dear Natasha and Nick,

Greetings once again! Now, I understand that you may have some concerns about your career opportunities, whether you have a bachelor's (BE), master's (ME), or doctor's (PhD) degree in engineering, or other equivalent designatory letters. From what I gather, you're not asking about which particular specialties of engineering exist, like aerospace, biomedical, chemical, civil, mechanical, etc. But, your question is: Where do engineers work?

Not surprisingly, this is one of the most common concerns I've heard from those thinking about a potential career in engineering, as well as those in the midst of their education and training in engineering but who are trying to plan for the future. If there's only a limited future in engineering, then why do it? And so, I want to address this extremely important topic.

Let's keep in mind that all sorts of nontechnological factors, like cultural, economic, environmental, and political circumstances, can certainly influence some of the details of what I describe below. Also, you'll notice that there is definitely some overlap in the possible career paths which may allow you to switch paths later on, so you should never feel like you are stuck forever. Now, although employers who hire engineers can be classified in different ways, I'd like to categorize your possible workplaces as either the university, industry, or government. Let's take a look at each of these (see Figure 12.1).

University Jobs for Engineers

There are usually 6 main career options for engineers in a university. Similar options can also be found in some technical schools or colleges. The position titles and other terminology that I use here, however, may vary somewhat depending on the institution or country.

The *tenured professor (regular stream)* is the most commonly-found and highly-prized engineering career at a university; this is what people usually mean by engineering "professor". The word "tenured" means this is a permanent and full-time position from which the professor cannot be dismissed, unless they commit a criminal act or seriously breach university policy; it's a job for life until retirement!

DOI: 10.1201/9781003193081-12

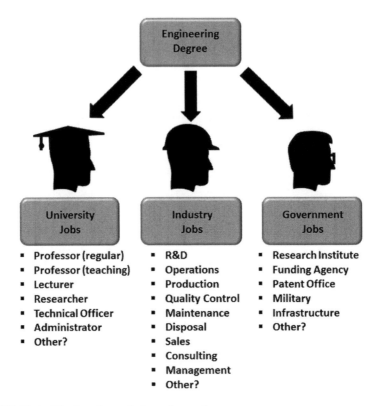

FIGURE 12.1 The 3 main workplaces for engineers.

To get tenured, an engineer with a recent PhD degree is often hired under contract as a junior professor, but then has to prove their worth by giving an outstanding performance over several years; otherwise, they lose the position at the end of the contract. Or, a tenured professor with a PhD at one university takes a position at another university, while retaining their tenured status.

Either way, junior and tenured professors both usually devote the same amount of time to 3 key areas. About 40% of their time is spent on teaching courses, which involves giving lectures, grading assignments and exams, and developing new curricula. Another 40% is focused on doing research, which involves writing applications to government, industry, or organizations to get funding or infrastructure support, supervising ME and/or PhD student thesis projects, publishing research findings in peer-reviewed journals, and presenting research findings at conferences. The final 20% is given to service to the home department or the university by active participation on various committees and task groups.

The *tenured professor (teaching-only stream)* is another possible career option, but only at some universities. This is similar to the tenured professor (regular stream) above, except the focus is on courses only. So, getting tenured depends on showing excellent performance in 2 key areas only, namely, teaching and service; there is no research component. To compensate for the extra free time, the professor will develop and teach extra courses, but may or may not need to do more service. And, depending on the courses being taught, this type of professor might only require an ME degree.

The *lecturer* has an ME or PhD degree and is hired only to teach and/or develop courses. They have no research or service duties. They may be hired permanently or be under contract for a period of time. Of course, any contracts may or may not be renewed, depending on their performance and the needs of the hiring department. The role may require full-time or part-time hours.

The *researcher* has a BE, ME, or PhD degree and is hired only to conduct research. They have no teaching or service duties. As above, they may be hired permanently or be under a renewable or one-time contract. The role may require full-time or part-time hours.

The *technical officer* has a BE, ME, or PhD degree and is hired to support the technological needs of a department or beyond. Their role may be to run student lab exercises, develop new computer software, do machine shop fabrication, help design test setups for professors, etc. There are no teaching, research, or service duties. As above, they may be hired permanently or be under a renewable or one-time contract. The role may require full-time or part-time hours.

The *administrator* has a BE, ME, or PhD and works on organizational aspects of a department or the university, rather than engineering per se. Different roles may be permanent or contract-based and require full-time or part-time hours. Junior positions open to those with a BE, ME, or PhD degree include working in the patent development office, health and safety office, etc. Senior positions open to a tenured professor with a PhD in engineering include the chair of an engineering department, the dean of an engineering school or faculty, the vice president of an administrative department (e.g., public relations, fundraising, personnel, research, etc.), and the president of the university.

I've known many people over the years who've had various types of engineering jobs at universities like the ones I just described. For instance, a number of my friends are tenured professors of engineering. This is probably the most prized job at a university, so I'll talk about it here as an example. These professors don't necessarily have a typical work day or work week because that depends on what's going on. But, I've noticed a couple of trends in these friends' activities. They're really busy with teaching courses during the fall and winter months. Yet, things slow down in the summer months because the school year is over and all the undergraduate students are gone. So, during the summer these professors take some well-deserved vacation time that's often a month long. And then they seem to focus better on research because the ME and PhD students they supervise and employ are still on campus in the summer. Also, some of these professors are involved in the same service work to their department over the years by sitting on exactly the same committee. But, others by necessity or choice seem to do different types of service work.

All these professors generally enjoy teaching, so they can shape the minds of the next generation of engineers. And they're very passionate about their research activities in developing new concepts and technologies. But, they all dislike the rivalries and tensions that sometimes arise between professors or between professors and the university's administration. When I was a student engineer, I remember one professor who didn't want his ME and PhD students talking about their research with another professor's ME and PhD students because of the competition that was going on between them. In another case, I knew several junior professors whose tenure status was either delayed or denied by the department's administration for political or personal reasons, even though these junior professors far outperformed some of the tenured professors in their teaching and/or research. Sad, but true! So, if you think you want to be a professor of engineering—or you want to work at another university job that I discussed above—just be aware that there are many rewards and responsibilities, but there are also risks.

Industry Jobs for Engineers

There are usually 9 potential career options for engineers in an industry, which may involve a small company, a large national or international corporation, a non-profit organization, a research institute, a consulting firm, and so on. Depending on the circumstances, each career below may have different titles, university degree requirements (i.e., BE, ME, or PhD degree), permanent or contract positions, and full-time or part-time hours.

The *R&D engineer* conducts research and development (thus, it's called R&D) of a new device, structure, software, process, etc., using analytical/mathematical modeling, computational analysis, experimental testing, on-site field observations, and/or back-of-the-envelope calculations.

The *operations engineer* uses analytical/mathematical modeling, computational analysis, experimental testing, on-site field observations, and/or back-of-the-envelope calculations to plan and run the day-to-day functioning of equipment, machines, processes, etc., within the company.

The *production engineer* uses assembly, fabrication, programming, and/or refining methods to build a device, structure, software, process, etc., the company needs for itself or to sell.

The *quality control engineer* uses computational analysis, testing equipment, and/or observation to ensure the new device, structure, software, process, etc., meets technical standards during or after production before it goes to the customer or client.

The *maintenance engineer* uses computational analysis, testing equipment, and/or observation to ensure the device, structure, software, process, etc., works properly and then fixes or replaces it if needed.

The *disposal engineer* ensures that outdated and/or irreparable devices, machines, structures, etc., are stored, discarded, disassembled, and/or recycled in a safe, legal, ethical, and effective way. This can also involve appropriately disposing toxic materials and waste products.

The *sales engineer* interacts with current or potential clients and customers to promote, demonstrate, and sell them a device, structure, software, process, etc., that the company makes.

The *consulting engineer* offers expert advice to companies, organizations, and governments that want to design, fabricate, purchase, fix, etc., any devices, structures, software, etc.

The *management engineer* is a middle or upper administrative role that involves supervising a department of the company, like R&D, production, maintenance, sales, etc., or the company itself.

Let me give you a real-life example of an engineering job in any industry by providing some insights into my own mechanical/biomedical engineering career. For many years, I've been in charge of several research labs and groups in various hospital settings in the healthcare industry. My main goal has been to develop new techniques and technologies for orthopedic surgery applications. And my additional aim has been to publish scholarly articles based on my research findings in recognized journals in order to contribute to my field and improve the healthcare industry as a whole. So, if I had to categorize myself at all, I'd have to say I've primarily been a senior R&D engineer. But, when I look back at what I've actually done, I realize I've also functioned in all of the other industry roles I described above.

For instance, I was like an operations engineer whenever I planned and directed the day-to-day functioning of equipment and machines in my lab. I was like a production engineer whenever I fabricated and assembled apparatuses for various experiments. I was like a quality control engineer whenever I assessed the quality of the fiber-reinforced composite materials I developed for use as bone fracture implants. I was like a maintenance engineer whenever I inspected and repaired equipment in my lab that wasn't working properly. I was like a disposal engineer whenever I stored, discarded, or recycled electro-mechanical devices and biohazardous materials. I was like a sales engineer whenever I wrote research proposals to convince funding agencies to support my work. I was like a consulting engineer whenever I advised orthopedic surgeons and engineers about designing, repairing, and purchasing equipment. I was like a management engineer whenever I supervised projects, activities, personnel, and budgets for my lab. I learned how to do many of these tasks informally and in my moment of need. My point here is that an official job title may or may not fully describe everything an engineer does in their career in whatever industry. So, if you want to be an engineer in industry, make sure you're ready to be a "jack of all trades" who's always learning new knowledge and skills for any new tasks that come your way.

Government Jobs for Engineers

There are usually 5 government branches that hire engineers, as listed in the following discussion. Of course, the existence, the number, and the interaction between these branches will vary from country to country, as well as from local to regional to national levels. As mentioned, different circumstances can also influence position titles, university degree requirements (i.e., BE, ME, or PhD degree), permanent or contract positions, and full-time or part-time hours for these jobs.

The *research institute* is concerned about the R&D of new devices, structures, software, processes, and so on, that can benefit society. The implementation of these new technologies can boost the communication, economy, healthcare, security, etc., of that society. Common positions for an engineer in a government research institute are R&D, operations, production, consulting, and management engineers.

The *funding agency* uses government funds to create research grant programs that support the R&D work of university and industry researchers. As mentioned earlier, the realization of new technologies coming from university and industry settings can also generate many possible spin-off benefits to society at large. The most common position for an engineer in a government funding agency is administration.

The *patent office* provides legal recognition, conceptual protection, and financial security for novel technologies developed by individuals, companies, and organizations primarily within the host country. However, some countries offer patents that are recognized in other countries too. The most common role for an engineer in a government patent office is administration.

The *military* performs its own R&D to create new offensive and defensive technologies for its country's security, peacekeeping, and battlefield goals. The military, however, may hire university and industry researchers to do this work for or with them. Some of these technologies can also be sold to other nations to boost alliances or for other strategic reasons. Common roles for an engineer in a government military division are R&D, operations, production, maintenance, and management engineers.

The *infrastructure division* focuses on producing and maintaining any infrastructure owned by a local, regional, or national government. This can include things like buildings, buses, power lines, railways, roadways, satellites, trains, waterways, and so forth. Common roles for an engineer in a government infrastructure division are operations, production, maintenance, and management engineers.

As a case in point, I'd like to tell you about my friend who works as an engineer for a government agency focused on R&D and consulting. This person has several responsibilities. They design experiments that will be carried out by other scientists and engineers. They work with other technical staff from similar government agencies around the world to coordinate their efforts and sometimes even travel abroad to provide technical assistance. And they write technical manuals about how to run equipment that has been developed in-house for future users. But, apart from these technical duties, my friend tells me that the most important abilities an engineer needs to have are good verbal and written communication skills, being able to work well as part of a team, and personal time management.

Moreover, from my friend's perspective and experience, being an engineer working for a government agency has several advantages and disadvantages. The government provides 100% of the funding to support the agency, thus, an engineer has good job security and can potentially be employed for 20 to 30 years. Also, a government engineer's salary is often comparable to engineers doing similar work in any private industry, although this can vary greatly depending on the size of the company. And a government agency supports its engineers who want to be ambassadors that speak to high schools, universities, and community organizations to promote engineering and recruit potential new employees. However, one major drawback is that getting permission for anything (e.g., projects, spending, traveling) usually involves a complicated multi-step approval process that often slows down the actual work of engineering. Another downside is that the government agency—not the engineer—automatically owns the intellectual property (e.g., patent, copyright, trademark, etc.) of any new idea or product that the engineer might develop while they are still employed by the government agency.

So, What's the "Take Home" Message of This Letter?

My main intention here was to let you know that there are many potential career paths you can follow to establish yourself as an engineer. The university, industry, and government are the 3 main sectors in society that hire engineers. And, perhaps surprisingly, many of the same engineering roles can be found in each sector and, thus, changing career paths is maybe easier than you think. Although there might be some things I've missed in this letter, I do hope it has been somewhat helpful. Let's stay in touch.

Onward and upward,

R.Z.

Letter 13 Start Your Engineering Career: How to Get a Good Job

Dear Nick and Natasha,

I feel fortunate that I can spend time thinking about your questions on engineering as a career. To be frank, this process has at times challenged me to re-evaluate my own attitudes and actions, which can be mildly uncomfortable. But, mostly it's been a real pleasure to engage the topic. In this dispatch, the perennial problem of how to acquire a good engineering job is the subject of inquiry. There are probably as many different approaches to this as there are grains of sand on the beach. Every gainfully employed engineer has their own story about the path they traveled to find their job. However, many of these stories have things in common, which will be my focus here. Much of this content is based on my own experiences, the experiences of my colleagues, and the various resources created by experts in finding a job.

Also, because some of this job search advice may seem generic and applicable to any field of endeavor, I'm going to sprinkle in some examples from my own engineering career, so it will be more relevant to you. Now, I don't claim that if someone reads this letter and puts everything into practice that this will guarantee them success in finding an engineering job. Yet, I believe and hope that the ideas here can help, at the very least, to hone a person's job hunting skills and strategy. A major caveat, of course, is that there are many things that are often beyond our control that could influence our ability to get a good job, like global economic changes, national political circumstances, regional ecological upheaval, company bankruptcies, personal health crises, and so forth. With that said, let's get to it.

What's Your Engineering Dream Job?

This is the first and most important question to ask because the rest of the job search process flows after it (see Figure 13.1). Why settle for something subpar, when you can have the best? Everybody's idea of a dream job is going to be different, based on their personality, circumstances, family commitments, stage in life, worldview, and a whole host of other things that are too numerous to list here. And, of course, a person's concept of their ideal job may actually change over time. So, take some time to think about what it is you really want and what factors need to be considered. Write down your ideas, or even record your thoughts on audio or video.

DOI: 10.1201/9781003193081-13

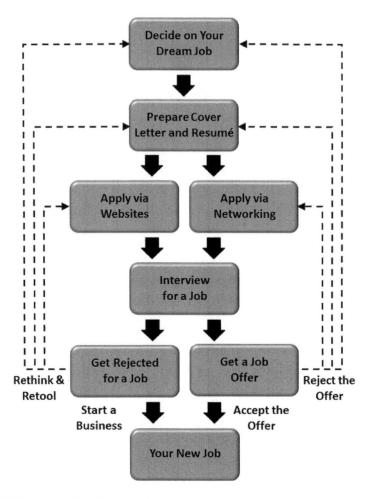

FIGURE 13.1 Searching for an engineering job.

What are important factors that would describe your perfect engineering job? Here are some possible criteria to reflect upon, but I'm sure there are others.

Work duties. Are there particular tasks, techniques, and technologies with which you have a lot of experience and/or that would suit you best? Do you have any health and safety concerns about the job?

Work terms. Are you looking for full-time or part-time hours, and would a lot of overtime hours be acceptable? Would a limited contract be okay, or will you only accept a permanent position?

Financial benefits. What kind of base salary, extra benefits, and retirement package are you looking for? What is a typical salary for someone with your level of education and experience?

Employer type. Do you want to work in a small company, which may have a more relaxed atmosphere but offers limited advancement opportunities? Or would a large corporation be preferred, which may have a fast-paced atmosphere, although it may have more chances for promotion? Is it a relatively new and inexperienced company, or is it an old and well-established organization?

Geography. What towns, cities, or countries would you be willing to work in? Have you considered the weather conditions, political situations, cultural practices, or commercial amenities in that location? Do you have any friends or family that already live there?

Social aspects. Do you have any family, friends, or dependents that your decision might affect positively or negatively? Should they be involved in your decision-making process?

Personal advice. Is there anyone you know whose advice you can get who has a similar kind of job, who works in that company, who lives in that city, or who has comparable family circumstances?

Writing a Cover Letter and Resumé

A typical engineering job opening will often attract dozens and sometimes hundreds of applicants. As such, the employer or hiring team will get swamped with many cover letters and resumés that they'll initially just skim over briefly to exclude weak applications. Then, they'll probably carefully read the cover letters and resumés only of those on the short list of applicants who are more seriously being considered for the job. Some employers, especially universities that are hiring engineering professors, will require additional information in the application package, such as sample academic publications, a teaching statement, a research statement, etc. So, you as the job applicant need to accomplish 2 important things in the application. You should show your uniqueness—in a positive way, of course—from all other job seekers. Then, you should try to persuade the employer that your education, experience, knowledge, and skill are the best match for what the employer wants and needs. Remember that your cover letter and resumé may need to be slightly adjusted for each specific job application. Here are some tips on cover letters and resumés.

Cover Letter

A good cover letter is very important. You're introducing yourself and your application. Employers will usually read this before your resumé or any supplemental documents. Here are my top 10 tips.

1. Keep it short. A cover letter should be 1 page maximum for industry or government job applications, or at most 2 pages for university professor jobs.
2. Make the greeting specific. The opening salutation should address a specific person by name and formal title, if you know it. Otherwise, you will need to greet the search or hiring committee.
3. State your objective. In the first line, state clearly for which particular job you are applying. Don't say you're applying for any job that's available, since it looks like you're not sure what you want.

4. Write a catchy next line. Indicate you're applying because you're the best candidate for that specific job. Then proceed to explain the exact reasons why this is the case, such as how your education, experience, qualifications, skills, etc., match the specific job profile.

5. Use formal and professional language. Don't write "I've" but "I have." Try to stay away from using slang expressions or idioms. It's also best not to attempt any humor in the letter.

6. Make sure the letter is clear, concise, and positive. Keep your sentences crisp and reasonably short. Try to avoid complaining about your prior workplace or about your bad current financial situation.

7. Use bold or italic type for keywords and phrases. Do this sparingly for specific things that you want to highlight, so the letter visually appears to have a logical structure.

8. Remember the details. Don't forget to add the date, the recipient's name, title, and contact information at the top, your full contact information, and your signature.

9. Double-check the letter. Make sure the spelling, grammar, font size (11 or 12 is typical), and page margins (1 inch is typical) are correct.

10. Get some feedback. Let a trusted friend or colleague read it and give constructive criticism. Get professional help from a job search agency or coach, if possible.

Resumé or CV

Your resumé or CV (i.e., curriculum vitae) summarizes your educational qualifications and professional experience. This is the core of your application. The employer will probably go through this in some detail. So, it deserves a lot of attention. Here are my top 10 tips.

1. Keep it short. Make the resumé 2–5 pages maximum, unless the employer asks for your full resumé. But, for an engineering job as a university professor, senior industry position, or government researcher, it is quite common to submit a full resumé which can be dozens of pages long. Thus, it might be a good idea to always keep an up-to-date version of a short and long resumé.

2. Identify yourself. State your name, educational degrees, current occupation (if you have one), and contact information at the very top of the first page.

3. Identify your goal. Give a strong and specific statement near the start about your career goal, tailoring it to the particular employer and job opening.

4. Summarize your career. Use the rest of the first page to summarize your education, work experience, and special skills that are really relevant to the particular job you're applying for. If possible, state them in the order they will appear in the rest of the resumé, almost like a book's table of contents.

5. Give the details. Use the rest of the resumé to provide the details of your education, work experience, teaching experience, special skills, publications, patents, awards, achievements, citizenship, etc. And also include some professional references at the end, such as prior employers, professors, or

work colleagues who can vouch for what a great engineer you are; make sure to get their permission to list them.

6. Keep it neutral. Don't put any information in your resumé that could potentially bias people against you. The truth is that we all have conscious or unconscious biases, including employers. You don't want them to reject you too quickly. So, I suggest you avoid including your age, gender, height, weight, photo, languages (unless relevant to the job), political affiliation, race, religion, etc.

7. Make it easy to read. Whenever possible, use bullet points and lists. This will make it easier for the employer to read. The last thing they want to do is read long sentences and essays.

8. Make it look good. For each major section of your resumé, use bold or larger font for the title. Don't be afraid to introduce color, such as a thin blue line that runs across the top or side of each page.

9. Double-check it. Ensure there are no spelling or grammar mistakes, and that the font size (11 or 12 is typical) and page margins (1 inch is typical) are appropriate.

10. Get some feedback. Ask one of your friends or colleagues to give you their opinion on your resumé. If needed, a professional job search agency or coach can be very helpful.

So, for example, as someone who's been involved in hiring biomedical engineering students and junior biomedical engineers, I can describe what I've looked for when I've reviewed cover letters and resumés. My main goal is to arrive at a shortlist of serious potential applicants, by excluding applications that are clearly not suitable as quickly as possible. Despite my best efforts in accurately describing the qualifications and duties in the job profiles that I've posted online, unqualified people have still sent me applications. The first step is to exclude those who are not qualified, such as those who don't have an educational background in engineering (e.g., physics or biology or medical graduates) and those who don't have the proper set of skills (e.g., no experience in computational modeling, mechanical stress testing, or fabrication techniques). The next step is to exclude those who have too many spelling and grammar errors or whose application is very disorganized. The reason is that writing skills are important for producing grant applications, thesis dissertations, progress reports, and journal articles. The final step is to exclude those whose last job was in a very senior engineering role, but who are now applying to me for a very junior engineering role; unless there is a really good reason for this, it makes me wonder whether that person left their prior job for dubious reasons. This process usually helps exclude about 75% of applications, while the remaining 25% of applications are closely scrutinized and compared to each other based on job qualifications and duties in order to arrive at a final shortlist of applicants I will interview.

Applying for Jobs via Websites

Back in the "old days" people would often look for work by scouring job ads and classified sections of the newspaper. Some things have changed, but others haven't.

These days, the equivalent of the newspaper is probably the internet. The problem is that the internet is so huge that it can be intimidating and, in fact, it may be hard to know exactly where to start. The plus side of this is that there are many websites out there that you can explore to find out what job openings exist. Because engineers are mostly employed by the university, industry, and government, those websites are probably the best place to begin, but there are other databases that should be explored.

I want to also briefly mention some things you probably should not do that would be a waste of your time and the employer's time. Don't apply if you are really not qualified for the job. Don't apply to an employer who doesn't have a job opening. Don't apply if you know the employer is hiring someone from within their organization, unless of course that includes you. Don't repeatedly contact the employer to see how the application process is going, especially if they clearly state in the job description that only suitable applicants will be contacted. Now, here are some suggestions for where to look and apply for jobs.

University websites. This is the best place to start looking for a position as an engineering professor, lecturer, researcher, or technical officer. Some universities have centralized job databases that you can explore and apply through, whereas others let the individual departments handle this. The same can be said for technical schools and community colleges.

Industry websites. This is a good place to start if you are looking for a position as an R&D, operations, production, maintenance, or sales engineer, or perhaps a role in management. Often, you can directly interact with the president or upper management of a small engineering business or organization, but you will often directly communicate with the HR (human resources) department for larger engineering companies or organizations. Make sure to double-check if they have branch offices, since the job opportunities may vary from branch office to branch office.

Government websites. Local, regional, or national government websites often have job databases that you can search for a variety of engineering jobs that are similar to the ones offered in industry. Also, you will probably be dealing with the HR departments directly, rather than the people with whom you would actually be working. Sometimes you can be automatically informed of new job openings by joining their email list, if they have one.

Professional engineering society websites. One of the advantages of being an official member of a recognized professional engineering society for a certain specialty (e.g., aerospace, civil, electrical, mechanical, software, etc.) is that they often provide information on job openings and even offer to help you find employment. But, they may not offer such services to non-members. Sometimes you can be automatically informed of new job openings by joining their email list, if they have one.

Research network websites. University engineering professors and other research engineers primarily join such online groups to exchange data, ideas, and journal articles with their peers. However, another benefit is that some of these groups provide lists of job opportunities, or at the very least you are able to post a message asking if anyone knows of a job opening. Sometimes you can be automatically informed of new job openings by joining their email list, if they have one.

General job websites. These extremely popular websites are probably the most common way that people look for gainful employment and that employers advertise their job openings. These websites let you input keywords to help narrow the search, such as engineering specialty, geographic location, salary range, and so on. They are usually free to use for general purpose job hunting, although some may require a fee for advanced services like reviewing your resumé or coaching you before an interview. Usually, you can be automatically informed of new job openings by joining their email list.

Social media websites. Professionally oriented websites often require you to be a member to formally search for job openings, although as a member you can also informally ask your "friends" or "contacts" if they know of any opportunities. Socially oriented websites may not have job search capabilities, but you can also make informal inquiries of your "friends" or "contacts" about engineering jobs.

Independent email lists. You can also join an engineering email list that sends you regular updates on general topics of interest, upcoming academic or industry conferences, and job openings in university, industry, or government. These are often run by volunteers as a service to their peers, rather than run by universities, industry, government, or professional engineering societies.

Recruiting agencies. Some engineering companies hire recruiting agencies to do the initial task of evaluating the many job applications they receive and/or directly contacting potential new employees who may, in fact, already be working at another engineering company. The agency whittles down potential new employees to a shortlist of the best candidates, which they pass on to the engineering company. So, the engineering company doesn't have any direct contact with you until you do a formal job interview. Also, if you do get a job this way, it's possible that it may only be a contract job for a limited period of time, thus, it often has a lower salary, fewer benefits, and no job security. But, if you perform well and the company has an opening, then the contract job could be converted into a full-time permanent job.

To illustrate the point, I was once in charge of seeking out and hiring a junior engineer who could run the daily operations of my team's biomedical engineering research lab. I wrote a simple 1-page job profile, got it approved by my colleagues on the team, and then proceeded to upload it to a well-known general job website and an independent email list. This did not cost me anything. I did not do any other advertising of the job. To my surprise, within about 2 weeks, I received almost 120 applications from all over the globe by email. And, from this group, a suitable person was eventually hired. This highlighted 2 things in my mind. First, engineering employers can widely advertise their jobs by making strategic use of websites with minimal time, effort, and cost. Second, engineers looking for a job through website searches should know that the job market is highly competitive with many people applying for the same position, so they need to make sure they stand out from the crowd (in a good way, of course).

Applying for Jobs via Networking

Back in the "old days," job seekers often asked their friends and acquaintances to recommend them to their boss. There was some wisdom in this. And the truth is, social networking is probably still one of the most effective methods for landing a job.

There is good scientific evidence for this based on "social network theory." In essence, the entire world is like one huge social network composed of a vast array of interlocking networks of networks, some small and some large. This entire global social network is held together by "super connector" individuals who have the largest number of links to other individuals and networks. It is these "super connectors" that are the most useful to us in finding a job by referral. In other words, don't only look to your family or closest friends for help finding a job, since they probably run in the same or similar social circles that you do. Rather, mainly look to those people in your social network that may be "super connectors" because they have access to people you don't, even if they are merely distant acquaintances you rarely communicate with. So, the friend of a friend of a friend, may actually be the best person to help you find a job.

Another way of thinking about this is through the 80/20 rule (also called Pareto's Principle), which also has established scientific evidence for it. It states that about 80% of your successful results come from about 20% of the things you spend your time, effort, and resources on. So, this means that only about 20% of your social contacts will actually be practically useful to you in finding a job, while the other 80% of your social contacts will be ineffective. If used wisely and deliberately, it is these social networking efforts that can potentially help you get a volunteer position, an internship, a part-time job, a contract job, or even a full-time permanent job as an engineer. But, you'll probably still need to give your cover letter and resumé to your potential employer, and then do an interview before you officially get the job.

As an engineer, you may be interested to know that these social networks can be modeled and predicted mathematically using the power law. Simply put, the power law is $N = an^b$ where N is the number of social links that a particular person has, n is the number of people that also have N social links, and a and b are fixed numerical constants that are unique to that social network. This can also be applied to other manmade and natural networks like ecosystems, chemical interactions, brain cell connections, website linkages, etc.

For instance, when I did my PhD (i.e., doctoral) degree and post-doctoral fellowship in mechanical/biomedical engineering, I met a lot of professors and fellow master's and PhD students at university. After graduating, over the next few years, I kept in touch with some of them occasionally, but most weren't close personal friends. At that time, I was really struggling to earn a living through a few short-lived contracts for medical writing and some part-time lecturing at another university. Unknown to me, one of my prior university acquaintances landed a full-time permanent engineering job running a biomedical engineering research lab in a large urban hospital in a nearby city. This person was also putting the finishing touches on their master's degree. Their supervising professor—who also supervised me for my PhD studies—suggested they contact me to provide some feedback on their master's dissertation. So, we met a few times to discuss their master's topic. Then, this engineer asked if I'd be interested in volunteering at their research lab; I said yes without hesitation. After a few weeks, amazingly, this turned into a part-time contract job for me. A few weeks later, this engineer told me they were soon going to quit their job to take another job in another country, and then they asked if I'd be interested in replacing them on a full-time permanent basis; I quickly said yes. After providing my resumé and doing a formal interview with the boss, 3 months later I was the full-time permanent manager of that same biomedical engineering research lab. All this happened because of the power of social networking!

Interviewing for an Engineering Job

This is probably something that most job hunters don't like to do. Whether it's by telephone, video conferencing, or in-person, it can be intimidating. And it feels like we're being judged and evaluated which, of course, we are. Now, some people are naturally shy and nervous, and others are naturally gregarious and carefree. But, one type of personality will not necessarily do better than another in an interview, although this can be a factor. Rather, I think good preparation beforehand is just as, if not more, important. So, here are some practical tips to keep in mind.

Know the format. Is it going to occur by telephone, internet, or in-person? If it's by telephone, then especially make sure you have some water close by to keep your voice lubricated. If it's by video, then especially make sure your appearance is optimal in grooming and apparel, the camera shows your whole face, and the background behind you looks neat and professional. If it's in-person, then especially make sure your grooming and apparel are appropriate, your body language is relaxed but formal, that you don't make any annoying gestures over and over again, and that you're not wearing strong soaps, perfumes, or colognes that can be distracting. An interview is not the time to experiment with fragrances and fashions.

Know the time and place. If possible a few days before an in-person interview, take a trip to the actual location, and even room, where it will occur. This will help you know exactly how long it takes to get there in order to arrive on time for the interview. And, this will help you know the atmosphere of the room in order to anticipate elements like lighting level, noise level, room temperature, washroom location, etc.

Practice your answers. Write down any questions that you might be asked by the employer, as well as your best possible answers. Then, practice giving your answers out loud, to make sure they are clear, concise, and honest. Maybe do a practice session with a friend or colleague, to make sure you answer the questions well and make any adjustments if needed. Obviously, if you are expected to give a short presentation to the employer, make sure to follow any formatting requirements and have some time to practice a lot beforehand, perhaps even getting a friend or colleague's feedback.

Practice your questions. Similarly, write down any questions you have for the employer. Almost always, they will give you a chance in the interview to ask them something. I recommend at most 2 or maybe 3 questions. Make one of the questions philosophical, such as how the employer describes the vision of the company or why this particular job has become available; this shows you want to understand the big picture. Make one of the questions practical, such as the amount of salary or the number of people who will be on your team; this shows you are a hands-on person too.

Talk only 50% of the time. From my experience and from what the experts have said, the best interviews are those in which the employer and the job applicant engage in a real dialogue. Each of them only talks about 50% of the time. So, if you have a tendency to either talk too little or talk too much, then please be careful about this and adjust things accordingly in the interview.

Bring your notes. If possible and appropriate, bring a small notepad, file folder, 3-ring binder, or electronic device which has your cover letter, resumé, practice answers, and practice questions. The employer may want to ask you something specific about your cover letter or resumé, so you want to appear prepared by quickly referring to your own copy. Also, a quick glance at your practice answers and questions can trigger your memory, so you can talk effectively when it's your turn.

Relax, you're the expert on "you." Remember that most of the interview is going to involve the employer asking questions about your experience, education, knowledge, skill, and so on. Who knows more about you than you? You are the real expert on you. The employer can't possibly ask you a question about you that you don't already know. So, relax.

Take a break the day before. If possible, take a break from preparing for the interview the day before the interview. Don't be over-prepared so much that you become tired, confused, or nervous. A string stretched too tightly will eventually snap. I suggest you do something relaxing and fun the day before to take your mind off the interview.

Say thank you afterward. Of course, right after the interview, thank the employer for their time and effort, as well as giving you the opportunity to apply. That's common sense. But, it's expected and appropriate to send another very brief thank you a few days after the interview. It will remind the employer about you. And, it shows you are really interested in the job. But, I recommend doing this only by email, since a telephone call, video message, or in-person visit may seem too intrusive.

At this point, I'd like to share my experience as a senior research engineer in hiring junior engineers, master's or PhD students, or post-doctoral fellows for biomedical engineering jobs. I want to give you an idea of what a typical engineering job interview might involve. I usually do a focused 30 minute telephone, internet, or in-person interview of the applicant. I start by thanking them for their interest and application. Then, I briefly describe the job duties, the workplace facilities, the other people they will work with, the salary and benefits, and the approximate timeline within which a hiring decision will be made. I also remind them there are other applicants being interviewed for the same job. Next, I ask a series of specific questions to see if they are really suitable for the job: why they are applying to this specific job; what experience they have with fabrication machinery; if they've done CAD and other computer modeling; what data analysis software they've used; what experience they have with mechanical stress testing; and if they have any concerns about working with biological specimens. I also want to know about their legal status in the country to determine if there will be any obstacles to hiring them. Of course, I make written notes to myself as we go through the interview. Then, I give them a chance to ask questions about the job. And, finally, we say our cordial goodbyes and end the interview. After the interview, I enter a numerical score for each response to each of my questions—as well as a score for their sincerity, reliability, maturity, and communication skills—in a large table that I use to compare the shortlist of applicants for the job.

Getting Rejected for an Engineering Job

Rejection never feels good. That's just our human psychology. But, it's part of the process of looking for an engineering job. So, don't take it too personally, and don't let it be too discouraging. Keep focused on your goal. There may be a whole host of reasons why you, along with probably many others, were rejected for that particular job. Remember, there could be tens or hundreds of other people who apply for the exact same job. Many of the reasons for rejection are beyond your control, while some reasons could be areas you can improve upon.

First, there may be other applicants with better skills, knowledge, and experience for that specific job, or who have a superior ability during interviews. Second, maybe the employer checked your other social media profiles and found something they didn't like about you; so, always be careful what personal information and photos you post online. Third, the employer may have already had a specific individual in mind that they planned to hire long before, but they were legally obligated to advertise a job opening because the law required a public hiring process; this is sometimes the case with engineering positions at universities and in governments. Fourth, unfortunately, there may be some conscious or unconscious bias the employer has against a job applicant's age, gender, race, etc., which they would never publicly state, but that still affects their choice of who they hire.

Regardless of the reasons you might get rejected, communicate politely with the employer, thank them for their time and effort, and wish them well with the person they decided to hire. But, don't probe them too much about why they didn't hire you. They don't legally have to state their reasons. You're probably not going to get a full answer from them anyway other than some vague response that they decided to go with a more suitable applicant. And you might appear to be pushy and, thus, leave a bad last impression about yourself. It's always best to leave a good last impression with the employer because you never know when you might cross paths again in the future.

If for some reason, you just can't seem to get any job offers after a long period of time and your financial situation is starting to look very stark, then maybe it's time to rethink your approach. Maybe it's time to rethink your idea of a dream job. Maybe it's time to get some expert help in rewriting your cover letter and resumé. Maybe you're looking in the wrong places, or not in enough places, for a job. Maybe you need to utilize your social network in a new way to be more effective in helping you find a job. Maybe you need to overhaul your interview skills. Maybe find a part-time volunteer position or unpaid internship, which is still something that can be added to strengthen your resumé, while you keep looking for a suitable full-time job. Maybe start a small engineering business with a few other friends who might be in the same situation, or perhaps become a freelance engineering consultant.

For example, I clearly remember being rejected for a particular engineering job. I sent in my well-prepared application by email, did a telephone interview, and was then invited to do a formal in-person interview with the hiring team. I realized I must have done well so far because I knew that hundreds of people usually apply for that type of engineering job. Then, I prepared well for the interview by anticipating the kinds of questions they might ask, writing down my answers on a sheet of paper, practicing my verbal answers many times, and preparing a short audiovisual presentation at the employer's request. On the day of the interview, I met and had meaningful one-on-one conversations with several key people, my presentation was well-received as confirmed by several people's comments to me afterward, and my interview with the hiring committee went smoothly as far as I could tell. The hiring committee even told me that my resumé was very impressive. I honestly thought I would get that job. To my surprise, a few weeks later I received a polite, but clear, email telling me that I was not going to be hired. I didn't take the rejection personally. So, I responded by email, thanking them for their time and effort, telling them I was disappointed but I understood this was part of the process, and wishing them luck in the future. Why was I rejected? I will never really know, nor do I personally need or want to know. But, a few months later when I was browsing the employer's website, I discovered who they hired. My resumé was much better than that person's. But, they were several years younger than me, so maybe that was a factor. Maybe they performed better than me in the interview. And maybe there was a skill they had that I didn't. My point is that we never know why employers make certain decisions, not to take it personally, and just move on.

Getting Offered an Engineering Job

If you get a job offer, or even multiple job offers at about the same time, don't necessarily say yes right away. Rather, consider how well a job offer matches with your dream job, which is something we looked at earlier. It's not likely that every job offer will match your dream job perfectly point-by-point, but hopefully it will match most of your expectations. There's always going to be something in a job offer that you might not fully understand or like. So, in many of these cases, it's reasonable to accept the offer.

But, if you are feeling desperate, don't get tempted to take a truly bad job offer that comes along that's almost completely opposite to your dream job. If you do take a bad job offer, how long do you think you will really last in that job before you decide to quit? Now, I'm not saying it's always a bad idea to take a bad job offer. It may actually be a good idea to take a bad job in order to get some engineering experience that can be added to strengthen your resumé, while you secretly keep looking for a better job.

One tried-and-true method of comparing a job offer to your dream job, or for comparing 2 or more job offers to each other, is to create a checklist of factors or a table of pros/cons that are each assigned a numerical value to indicate their level of importance. Then, add up the numerical values of all the factors or pros/cons that you checked off for each job and compare their totals. Such an exercise can help you be as specific and objective as possible, rather than just randomly making decisions.

In any case, if you accept a job, obviously it's really easy to be polite and grateful over the telephone, video meeting, email message, or paper letter. That's not really a concern. But, if you decide to reject a job offer, don't just remain silent and not respond at all. Send a short, clear, and polite communication telling them about your decision and, if appropriate, the reason for it. This is just common courtesy. Also, the other reason is that you never know if you will cross paths in the future with that person, company, or organization, so make sure to leave a positive last impression on them. At this point, either rethink your idea of a dream job, in case you've been too picky. Or, just get back to tweaking your cover letter and resumé for each new particular job opening that fits your original dream job idea, and then proceed once more with the process of applying and interviewing.

I think it would be worth sharing a personal situation I had early in my engineering career in accepting and rejecting job offers. A few years after finishing my PhD in mechanical engineering with a specialization in biomedical applications, I secured a part-time contract job as a university lecturer for a few biomedical courses. There, I met some nice people, it was a good learning experience, it paid my bills, and it was something I added to my resumé to strengthen it; I did this for about a year. Then, my boss at the university offered me a renewed and improved, but still part-time, contract as a lecturer, whereby I would teach additional courses and earn more money. I gladly accepted this renewed job offer, which would start again a few months later when university classes resumed. However, a short time later through some accidental social networking that I described earlier in this letter, I wound up getting a second job offer for a full-time permanent biomedical engineering research job. After carefully considering both options, it was clear that I really needed to take the full-time permanent engineering job; it was in-line with my long-term career goals. So, I arranged to meet face-to-face with my university boss in their office and tell them politely that I had to withdraw from the course lecturer contract because, in the meantime, I found a better full-time permanent job. Of course, I felt bad about the situation and apologized to my boss because it would put extra stress on them to find a replacement for me. Fortunately, my boss was very kind, supportive, and understanding of my reasons for taking the other engineering job.

So, What's the "Take Home" Message of This Letter?

The core message here was that looking for and finding a good engineering job not only takes time and effort, but also there are tactics that can help you. And it is these tactics that I've tried to highlight in a logical and useful way. It's not enough just to have a good engineering education from a university, previous experience working as an engineer, or even connections on the inside. But, you need to know how to sell yourself and even make yourself stand out in a good way from the crowd of other applicants vying for the same job. I do hope this letter has given you pause to think about how to improve your efforts in finding, not just any old engineering job, but your dream job.

Good luck,

R.Z.

Letter 14 Keep Your Engineering Career: How to Make Your Boss Happy

Dear Natasha and Nick,

Hello to you from my computer keyboard. It's become a second home to me as you kindly allow me the privilege of writing this series of letters to you about engineering. In this missive, I'd like to concentrate on one of the most important relationships you'll ever have in your career as an engineer, that is, your relationship with your boss! They may be awesome. They may be horrid. Or, maybe, they're just average. Whatever the case is, an engineer's success and satisfaction will hinge, in large part, on making the boss happy. And this requires them not only to work hard, but also to work strategically.

The tips I want to give here may just seem like common sense, yet the challenge is to put these tips into action! Because these tips are certainly transferable to any area of endeavor—education, entertainment, family, politics, religion, sports—I want to make them relevant by giving some practical real-life examples to show how the tips can be applied by engineers in the workplace. Also, I'm sure that my suggestions below can be reorganized and expressed in different ways, but I hope they inspire you to continue learning about securing your job by making your boss happy.

Now, I'm definitely not saying that you need to make your boss happy even if they ask you to lie, cheat, steal, compromise on health and safety standards, etc., or if you somehow feel peer pressure by other employees to do so. But, I am encouraging you to try your best to make the boss happy within the boundaries of your own personal values, professional engineering standards, employer policies, and the laws of the land (see Figure 14.1).

Having said that, the tips I want to share can be summarized by an acronym that I've created called BE FAST. The first part of that is BE, which has to do with your boss and the company where you work, whether it's a university, an industry, or the government. The second part of that is FAST, which has to do with your own thoughts, attitudes, words, and actions. Let's examine each of these below.

B Stands for "Boss"

This means you study your boss carefully to find out who they are and what they're like. Then, you'll know how to anticipate their actions, needs, and wants and then respond to them effectively. Obviously, this will take some time to do properly, so that

DOI: 10.1201/9781003193081-14

FIGURE 14.1   Making the boss happy isn't always easy.

you can eventually implement your strategy of "managing up" or "managing your boss." So, start by asking yourself these kinds of questions. What inspires your boss—is it money, power, productivity, recognition, reputation, winning, or something else? What worries your boss—is it budgets, criticism, deadlines, failure, stress, technological problems, or another issue? What is your boss's management style—do they micromanage you just waiting for you to make a tiny mistake, keep you at arm's length without knowing exactly what you're doing, or something in between? When is your boss the most, and the least, available and approachable if you need to talk to them about an important matter? What triggers your boss to be in a really good or bad mood, or do they always seem to be in the same kind of mood no matter what happens? etc.

For example, I know an engineer who had a meeting with the boss soon after being hired to discuss exactly what the boss's expectations were for the new job. This was their way of studying the boss. After asking the boss some key questions, the boss clearly indicated that they didn't care how many hours the engineer worked each week, whether or not they worked on weekends, whether or not they worked from home from time to time, or even when the engineer showed up to work each day. The main issue for the boss was that the various tasks and projects got done accurately and on time; in other words, this boss was motivated by productivity and not by the clock. So, this engineer felt free to arrange their own schedule to suit their own personal needs regarding travel time to work, peak physical energy, meal times, social life, and so on, as long as they were productive. Fortunately, this engineer was a very self-motivated and organized person, so this arrangement worked well and the boss was always satisfied.

E Stands for "Explore"

This means you explore your company (e.g., university, industry, or government) carefully just like you'd study your boss. It's good to know how the whole system works. This will help you predict and effectively deal with any changes in goals, rules, structures, and workflows because this can affect your department, team, boss, and you. So, make sure to read the memos circulating around the office, workshop, or laboratory, the company newsletters sent to employees by email, the company policy statements and brochures given to employees when they started work, and so on.

But, also keep your ears tuned to any useful information that comes out during casual conversations around the lunch table, at a department social event, or walking from one part of the building to another, etc. It's also always a good idea to become familiar with the physical layout of your workplace by taking a look at floorplans and maps, walking around the facility or campus, and, if appropriate, introducing yourself to the heads of other departments to find out about what they do and perhaps even get a quick tour of their area. And last, but not least, it doesn't hurt to know something about the surrounding community where your facility or campus is located, just in case there are people, businesses, organizations, stores, and other resources that could benefit you at work.

So, for instance, early in my career as a junior engineer I became the day-to-day manager for a hospital-based biomedical engineering research lab. I soon realized I couldn't complete all my tasks and projects by relying only on the resources in the lab. I knew I had to explore and look outwards. I reached out to other departments in my hospital to do some specialized medical imaging of specimens for me that my lab was not equipped to do. I explored the wider community to discover an out-of-the way, but fully equipped, hardware store within a 5-minute walk from my lab. I connected with several engineering professors and technical officers at a university that was about a 15-minute walk from my lab, which resulted in many collaborations on projects that were published as research articles in peer-reviewed journals. And I linked up with several other biomedical engineers and orthopedic surgeons at other hospitals in the city to work together on several collaborative projects. All this was possible because I decided to explore.

F Stands for "Faithful"

This means that you keep your promises to your boss. If you say you're going to do this and that, by such and such a time, then do your absolute best to actually get it done. If you're able to achieve this on a consistent basis over a long period of time, then your reputation for faithfulness, trustworthiness, reliability, or whatever other word you want to use, will skyrocket in your boss's eyes. It will make your boss happy.

And it will better position you for potential promotions to roles of greater responsibility like team leadership, rewards like salary hikes, and even more personal satisfaction in your career. And, when you need to ask a special favor of your boss, they will likely be more open to it because of your proven track record of reliability. This doesn't mean that you can never afford to make a mistake in underestimating the time, energy, or resources needed to complete a task or project. If this happens once or twice, you can still recover from those mini failures by quickly getting back on track with your reputation of reliability. But, if you are consistently inconsistent in keeping your promises, then you can forget about all the benefits I just mentioned. In fact, if you get a reputation for always overpromising and underdelivering then your boss will, obviously, be unhappy and your very job may be at risk.

This has been true in my own career as a mechanical/biomedical engineer. One of my vital tasks has been to lead research teams to complete computational and experimental projects. Then, my duty has been to ensure that our team's efforts result in articles that are published in peer-reviewed academic journals. Because of my own mistakes in organizing my time, personnel, and resources, there have been a few projects I didn't complete, but this rarely occurred. However, the vast majority of the time I acquired a reputation for being faithful to fulfill my promises and produce concrete outcomes. And so, some of my colleagues gave me nicknames like "The Finisher" and "The Manuscript Monster."

A Stands for "Available"

This means you'll always make time in your schedule to participate fully in whatever task or project that your boss assigns to you. Under normal circumstances, obviously, your regular working hours at your workplace are dedicated to doing what your boss asks, and so this should not be a problem.

But, there can always be emergency situations that will challenge your availability. Maybe you're also working on another task or project that your boss's boss gave to you. Maybe a major deadline for a task or project is very close to the start of your scheduled vacation. Maybe your boss is asking for people to work late or on weekends until a critical task or project is done, but you have personal and family commitments during those times. How does a person deal with this? Quite frankly, it's not easy. You may need to talk to your boss about getting a deadline extension. Maybe ask another employee for their help. You might even consider changing your personal or family plans. The point is, do whatever needs to be done, so your boss knows that you are really available. And, finally, always be on time for appointments and meetings with your boss; in fact, if possible, show up early.

Now, I once had a research engineer who worked directly under my supervision who is a good illustration of scheduling their time to be available. There were always some projects that needed to be done by a certain upcoming deadline, so this engineer decided voluntarily to come into the office and lab after regular working hours and on several weekends to make sure the projects got done. Of course,

I encouraged this person to take a few days off from work after the projects were done, so that they could get some proper rest. This engineer also gladly participated in any team meetings that happened either before or after regular working hours in order to accommodate the challenging and conflicting schedules of the other team members. In these cases too, I encouraged this person to take some extra time off, so they wouldn't suffer exhaustion. In contrast, I know an engineer who was employed full-time, but simultaneously tried to do a part-time doctorate in engineering. Unfortunately, the time pressures on this person eventually proved to be too much to bear. So, they decided to withdraw from the doctoral program because they realized they could not be truly available to complete the degree requirements.

S Stands for "Skilled"

This means you put in the time and effort to become skilled at your job. You can do this in a number of ways. Have all the necessary information or knowledge to complete the task or project. Come prepared for all appointments and meetings with all the needed photos and graphs and data. Know how to use all the necessary computer software, fabrication machinery, testing instrumentation, and so on. And be familiar with the rules and policies of the team, department, and company.

Of course, any company that hires a new engineer knows that there is going to be some learning curve the new employee goes through before they'll perform at peak efficiency. And, to help speed up that development, all companies will train their new engineers either through an informal mentoring method or a formal training course. This is on-the-job training, and it's quite normal. However, once you go through that initial training process, you'll still have to learn new things along the way when the company takes on new policies, projects, techniques, or technologies. In these cases, it's really critical to become skilled in these new areas to keep your boss happy.

As a case in point, I've supervised a steady stream of research engineers and engineering students in my biomedical engineering labs who became highly skilled in a large variety of techniques and technologies. Either I arranged for a technical officer from the supplier to do a formal training session with my team, I personally trained my team, or I asked someone on my team already skilled to train the new person. These engineers and students always put in the necessary effort to become skilled in fabrication machinery (like drill presses, milling machines, lathes), testing equipment (like mechanical loading frames, optical 3D motion tracking systems, thermographic stress analysis cameras), and/or computer software (like CAD modeling software, finite element analysis software). And the result of this group of highly skilled engineers and students was that my team's research projects were routinely completed very quickly, efficiently, and accurately, and then published as articles in peer-reviewed academic journals.

T Stands for "Teachable"

This means that you are open to learning new things without complaining, resisting, or sabotaging. This could include adapting to changes in company policies or rules, learning how to use new software or equipment, taking on a different role on the team, adjusting your workflow to a newly announced project deadline, and so forth. Remember that you may not be the only one that has to accommodate the new situation. And your boss probably already has enough pressure on them to adapt to that same new thing, so they probably don't want to deal with someone who is undermining the process by constantly grumbling or, worse, deliberately undermining their authority. Needless to say, if you are that person, this will not make your boss happy with you. So, be teachable. Say yes, smile politely, and get back to work.

But, if you do have some genuine concern that the new thing, whatever it is, could endanger employee health and safety, create financial losses, introduce workflow inefficiencies, etc., then talk to your boss. However, make sure to suggest an alternative option for accomplishing the same task or project. These other options should not only address your original concerns, but also allow the task or project to be completed in a timely and effective manner. Remember to present your alternative option in a humble and diplomatic way, so that your boss doesn't feel stupid, incompetent, or criticized, or that you are a rival to their authority. The best-case scenario is that your alternative option may even make your boss look good in front of others because they were willing to implement your ideas. If you can accomplish that, then your boss may actually be grateful and see you as a valuable ally in the workplace.

I saw this quality perfectly exemplified in a former medical doctor who enrolled at university to do a master's of science degree. But, what complicated matters was they wanted to do their research thesis on a biomedical engineering topic. And this person didn't have any background in biomedical engineering. So, they enrolled in a new individual study course on the topic, which required them to read almost 30 chapters or articles, write a 30-page report, give an in-person live presentation, and pass a final exam. They completed the course enthusiastically and successfully. Moreover, this student voluntarily spent many hours in informal discussions with a willing local expert who was not their official thesis supervisor, just to learn even more about biomedical engineering. This student wanted to do a superb job on their research thesis. And, indeed, that is what they did. The student completed their research thesis, graduated with their master's degree, and published an article on their findings in a respected peer-reviewed biomedical engineering journal. All this was possible because the student had a teachable attitude. In contrast, I know an engineer who, sadly, had an employee who resigned because they were not truly teachable as they did not want to learn how to use fabrication technology to create their own specimens for mechanical testing and data analysis. Rather, the employee only wanted to do the testing and analysis portions of the job.

So, What's the "Take Home" Message of This Letter?

The goal of this letter is to encourage you to develop a good working relationship with your boss. This is one of the most important relationships you'll have in your engineering career. It may be the difference between getting a promotion (or not), getting a salary increase (or not), and even keeping your job (or not). You want your boss to be happy with you. So, I've tried to offer some practical suggestions and real-world examples of how the ideas can be applied in the engineering workplace. And, if I may, I'd like to encourage you to further ponder these common-sense ideas, take practical steps to put them into action, and keep your eye out for other good resources on the topic.

With best wishes,

R.Z.

Letter 15 Advance Your Engineering Career: How to Climb Ladders and Cross Bridges

Dear Nick and Natasha,

I'm cheered by your openness and interest in my letters on engineering. Thank you kindly. It motivates me to put extra thought into them. In this one, I'm going to tackle the subject of advancing your career. Personally, however, I don't think there's anything wrong with being satisfied with your current job situation and, thus, having no great ambition to change career paths. There's a good deal of comfort in having a routine, knowing what to expect, and avoiding the stress that even comes with good changes.

I honestly think that being an "underachiever" is wonderful. By that I mean there's more to life than just our careers. There are other priorities we probably should put our time, effort, and resources into. There's a saying I once heard that goes something like, "Some people live to work, but other people work to live." There's some truth to this. At some point in many people's careers, they don't want to feed their ambitions anymore and simply want to coast along in their current job until retirement. If this is you, there's absolutely no shame in it. And so, there's probably no need to read this letter.

But, if you're naturally an ambitious person who's always looking for the next challenge or opportunity, or if you haven't reached that point yet in your engineering career whereby you want to coast along, then this letter is for you. In this communique, I'd like to discuss advancing one's career, thinking through the potential good and bad consequences, and making a practical action plan to achieve one's goals. Because this generic topic is applicable to many fields of endeavor, I'll also provide some real-world stories of how this has actually worked for engineers I've personally known.

Climbing Ladders and Crossing Bridges

I sometimes find metaphors useful in understanding and communicating ideas. Of course, metaphors are limited and don't necessarily take into account all the possible exceptions. Even so, let me suggest that there are primarily 2 ways of advancing one's career.

The first way to advance your career is to *climb ladders*. Perhaps you've heard the phrase "climbing the ladder of success on the way to the top" or something similar. It means that a person will move *vertically* from a lower rank to a higher rank position within the same or different company or organization. The usual motivation is some combination of more money, prestige, authority, and so on. And, in my view, if someone has the talent, opportunity, and initiative to do so, then I wish them all the success in the world, as long as they make the climb upwards without hurting themselves or others.

The second way to advance your career is to *cross bridges*. Perhaps you've heard the joke that asks, "Why did the chicken cross the road?" The answer is, "To get to the other side." It means that a person will move *horizontally* from one rank to another similar rank within the same or different company or organization. This could be considered more of an optimization, rather than an advance. The typical reasons could be a combination of higher salary, more desirable work duties, kinder boss, less stress, family reasons, and so forth. Again, in my opinion, if a person's priority is simply to shift from one job to another very similar job which may have some minor added benefits, then I genuinely hope they succeed.

To be sure, an engineer can advance, or optimize, their job over the course of their entire career by using both tactics. Sometimes they might shift vertically upward and at other times they might move horizontally across. But, there may even be situations when it's tactical to seek a temporary job demotion in order to position oneself for an eventual huge vertical move upward. The point is that there is no end to the number of ways of adjusting one's career, it just takes courage and creativity.

What Are the Pros and Cons?

One way of changing your career situation is not better than another. Everybody is different in personality, priorities, and circumstances. So, the trick is to really know ourselves, as the ancient Greek philosopher Socrates (470–399 BC) once said, and figure out what kind of job change we might be looking for. From my experience, ironically, I don't think most engineers are really systematic or comprehensive in thinking about making changes to their career. There are usually a few added benefits that attract them to a potential new job, and then they go for it. Unfortunately, sometimes this doesn't work out as well as they hoped, and they regret making the change. I'd like to suggest a checklist of things to ask ourselves before we try to climb a ladder or cross a bridge. This checklist can be more easily remembered by the acronym that I've crafted called PASSENGER.

P is for "position." This refers to the official rank within the company or organization hierarchy. Will the new job you are considering move you up or down in official position and title, or will you be at about the same rank as before?

A is for "authority." This refers to the type and scope of decision-making ability a job holds within it. Will the new job give you more, less, or the same amount of authority, power, and responsibility compared to what you had previously?

S is for "salary." This refers to the compensation package that a job offers, including base salary, benefits, vacation time, and retirement plan. Will the new job offer you more, less, or about the same salary, and is there a fixed salary scale or is salary a negotiable item?

S is for "stress." This refers to the psychological demands and physical difficulties that go along with a particular job. Will the new job bring you pressures on a daily basis that you can truly handle, as well as the occasional work crises that will likely occur?

E is for "engineering content." This refers to the amount of scientific and technological content of a job. Will the new job let you use all the engineering education and training you've received over the years, or will you be moving more and more into the business or management side of things?

N is for "network." This refers to the type and number of professional contacts that a job affords. Will the new job allow you to expand your network of professional contacts in a strategic way for your future career plans, will it make you more marketable in your field, or will it shrink those opportunities?

G is for "greater job security." This refers to the likelihood that a certain job will still exist in the long term based on the goals, financial strength, and societal relevance of a company or organization. Will the new job be a long-term place for your career, or will you need to start looking for work again soon?

E is for "extra accountability." This refers to the moral, legal, financial, and technological liability or culpability that a person with a job has to face. Will the new job make you personally more or less answerable to others for the failure of tasks and projects?

R is for "reputation." This refers to the prestige or standing that a job has within the company or organization and beyond. Will the new job enhance or diminish how others think of you in your specific workplace, in the engineering field, in your family, among your friends, or within society?

Make an Action Plan for Change

Once you've carefully considered the benefits and drawbacks, as mentioned in the earlier section, regarding a potential career shift, then it's time to make and write down an action plan (see Figure 15.1). Many people think that the vague ideas rolling around in their mind—for example, they want a higher salary or they'd like to have a higher rank in the company—are goals. They're not! Those are just notions. To turn mere notions into reality requires creating a concrete plan for achieving your goals, as backed by scientific studies and anecdotal evidence. This is absolutely vital. After all, there is some truth to that old proverb that says, "If you fail to plan, then you are planning to fail." Here are my suggested steps in doing so.

Step 1 is to *choose your goal(s).* Make sure it's specific, measurable, and realistic. And write it down. If needed, use the PASSENGER checklist to help you decide. For instance, ask yourself questions like these. Do I want a particular rank, role, or title in the company or organization? Do I want to have direct authority over specific tasks or projects? Do I want to reach a certain annual salary level? Do I want to work from home a certain number of hours each week, so I can bring down my stress? etc.

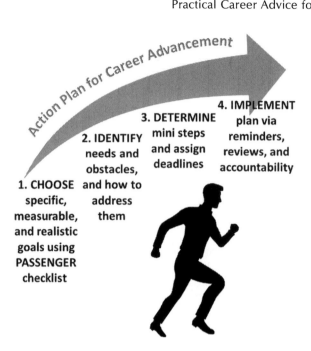

FIGURE 15.1 Create an action plan to advance your engineering career.

Step 2 is to *identify what you need to reach your goal(s).* Be specific here too. Right now, it's possible you just aren't qualified to advance your career the way you want. Maybe new "tools" need to be added to your "toolbox." Otherwise, this will be a perpetual problem. So, do you need to learn a new technical skill, acquire a new university degree, obtain a professional certification, learn from an experienced mentor, talk to a particular individual, make a geographic move, etc? Also, consider if and how you can address other obstacles that might get in your way, like personal bad attitudes or habits, rival applicants for the same job, alterations in employer policies, changes in societal laws, etc.

Step 3 is to *determine all the mini steps.* Make it super clear. Perhaps draw a road map or flow chart, or write down a checklist or table. Then, input all the individual mini steps that need to be taken that will eventually lead to the desired main goal(s). For each mini step, write down the exact timeframe or deadline, any people or resources you need, and any competition or obstacles that might be faced and how you'll overcome them. Don't be hasty. Take your time to think through this process.

Step 4 is to *implement the action plan.* Now it's time to put the plan into effect. Remind yourself of the plan by taping it to the wall or putting it on the computer screensaver. Review the plan frequently, say, daily or weekly. Focus on each mini step until they're each completed. Maybe give yourself a small reward each time you complete a mini step. Make small adjustments to the plan as you go along because you never know what new insights, information, opportunities, or obstacles may arise. Perhaps even ask a friend, family member, or trusted work colleague to keep you accountable in sticking to the plan.

Some True Stories of Engineering Careers

The following series of vignettes are from the careers of engineers I've known. You'll notice that some "climbed ladders," others "crossed bridges," and many did both in their efforts to advance their careers. I don't know what their personal motives were for each job change, how systematically they planned each advance or adjustment in their career, or how happy they ultimately were with each new adventure. But, that's not the point of these tales. I simply want to highlight the fact that each engineering career was unique and that it had its own twists and turns.

Engineer #1 completed a bachelor of engineering (BE) and an master of engineering (ME) degree. Then, the individual moved to another nation to enroll at university in another ME program, but because of their excellent grades after 1 year, they upgraded directly into a doctor of philosophy (PhD) engineering program. Next, they moved to another city and university for 1 year to do extra engineering training as a post-doctoral fellow (PDF), which was followed by yet another move to a different city and university for 1 more year as a PDF. Subsequently, they moved again to another city and university to take on a new role as a junior professor of engineering, but after 1 year for family reasons, they moved again to another country and university where they also got a job as a junior professor of engineering.

Engineer #2 graduated with a BE degree and then worked as a high school teacher for a year. Then, for a year this person was employed in industry as a quality control engineer. Next, they moved to another country and company for several years because they found work as a production engineer. After this, they moved yet again to another nation to obtain ME and PhD degrees in engineering, as well as extra training as a PDF at the same institution. Then, they moved to another university where they found employment as a junior professor of engineering. But, over the next few years, they were promoted several times to finally become a full tenured professor of engineering.

Engineer #3 finished their BE and ME at one institution but then moved to another university to do a PhD and work as a PDF for 2 years. This person moved again because they accepted a part-time role with a nonprofit charitable organization, but still did freelance engineering work on the side to supplement their income. Next, they taught some university courses for a year. But, then this individual found a full-time permanent position as a senior research engineer in industry, while becoming formally affiliated with a university so they could formally supervise the research projects of engineering students.

Engineer #4 did a BE at one university, then graduated with an ME from another university in the same nation. This person then obtained a full-time job at that same university working in administration as a patent officer dealing with science and technology. During their time of employment, they were also able to complete a master's degree in business administration and a master's degree in intellectual property law. This upgrade of their skill and knowledge helped them greatly with their duties as a patent officer.

Engineer #5 completed a BE in their home country, but then moved to another nation to study for their ME degree. Immediately after graduation, this person secured a role as an engineer in industry in charge of fabricating products for a company. After a few years, the company supported the engineer's desire to move back to their home country for family reasons. While employed with the same company, that individual set up an office and small testing lab in their home and communicated regularly with the head office.

Engineer #6 obtained a BE, worked in industry for a few years, and then did an ME. They moved to another nation to enroll in a PhD engineering program. Upon graduation, this person found a job as a senior technical officer and sales engineer working for their former PhD supervisor who had started a small spin-off company. A few years later, that small company was purchased by a major corporation. So the engineer was promoted to work in the head office of that corporation, which was located in another country.

Engineer #7 received their BE while working at contract jobs as a junior engineer when university classes were not in session. Then, after being a research engineer for a university professor, this person completed an ME. A big change then took place when they left engineering altogether for several decades to work for a nonprofit charitable organization. However, they eventually returned to engineering in a sales and consulting capacity for almost a decade before finally retiring from the workforce.

Engineer #8 completed a BE in their home country, then moved to another nation to finish an ME and PhD degree. Then, for a number of years this person worked in industry before returning to the university to become a professor of engineering. This individual gradually advanced their engineering career by becoming the chair of their department, then the dean of engineering, and then the vice president in charge of a major division of the university. Finally, they became the president of another university for a number of years before retiring from the workforce.

So, What's the "Take Home" Message of This Letter?

The primary idea I want to convey is that we should think carefully about the pros and cons of changing engineering jobs, either vertically or horizontally. If we're happy where we are, then maybe there's no need for change. After all, it's okay not to have grand ambitions and just bloom where we're planted. But, if we decide that a career change is something we really want and need, then I hope that my suggestions are instructional and my examples are inspirational.

Best regards,

R.Z.

Letter 16 Personal Organization Skills for Engineers: The "Clompass" Strategy

Dear Natasha and Nick,

How are you doing these days? I hope your response is positive. There's an old saying that states, "There's no time like the present." And so, it's time for another letter from me. Speaking of which, the topic of this missive is closely related to time. It's about managing your time and your priorities as an engineer. I guess it's really about managing yourself. I like to refer to this as personal organization skills.

Unfortunately, most engineers have only informally and haphazardly picked up their personal organization skills, but have never learned about it systematically via a course, book, video, and seminar, or even been mentored by an expert in this area. Yet, this is crucial for an effective and successful engineering career, since an engineer routinely deals with all sorts of people, tasks, projects, and timelines. If you make yourself more efficient, you will be more effective and valuable to your team, employer, client, and customer. And so, this is what motivates me to write about this topic.

I'll try to balance the rather generic nature of the strategies and tips I'll describe, by also relaying to you the true story of my own journey of how personal organization skills can improve an engineer's efficiency and effectiveness. I can honestly say that I wish an engineer with some know-how had shown me, or taught me, about how to manage my time and priorities when I went to school for engineering or during my engineering career.

But, like most engineers, I had to learn this on my own through trial-and-error, reading books, and often accidentally noticing what others did or didn't do. Although I'm certainly not an expert in this area—since I'm still learning—I'd like to distill everything I've learned about the topic in this letter. So, I hope this letter is useful and motivational to you.

Overall Strategy: The "Clompass"

The traditional "to do" checklist some people use to manage tasks is flawed because it doesn't require prioritization and it doesn't assign a timeline. In the workplace, this may be good enough for just a few days from time to time. Or if you generally don't have many responsibilities to deal with in your job. Or if you're always told exactly what to do, how to do it, and when to do it in your job. But, for an engineer who wants a satisfying and successful career, there's a better way.

DOI: 10.1201/9781003193081-16

Compass: Priority Tasks are Determined

Clock: Priority Tasks are Scheduled

FIGURE 16.1   The "clompass" strategy for personal organization.

I'm proposing what I call the "clompass" strategy (see Figure 16.1). This is a combination of the words "clock" and "compass." The compass part of this approach deals with assigning priority levels to tasks based on your short-term job duties (e.g., repair this machine, purchase a device from a supplier, calculate the stress on this component, etc.) and also your long-term career goals (e.g., what projects do you need to finish to position yourself for a promotion? what skills do you need to learn to get an increase in salary? etc.). Then, the clock part of this approach deals with assigning time blocks in your daily calendar to work only on the most important tasks, and then assigning final deadlines for those tasks.

Here's how this can work practically. Initially, you would organize your tasks into 4 categories, such as 1 (important and urgent), 2 (important and not urgent), 3 (unimportant and urgent), and 4 (unimportant and not urgent). Then, schedule specific time blocks in your daily calendar to only work on important tasks 1 and 2 until they are totally completed. Finally, for unimportant tasks 3 and 4, you either don't ever do them, work on them only if and when you finish important tasks 1 and 2, delegate them to someone else in your workplace, or outsource them to external departments or businesses.

General Tips

The overall "clompass" strategy is primarily about managing your priorities within certain time frames. But, there are many helpful hints that you can incorporate, or should at least be aware of, as you work the "clompass" approach more effectively. Of course, you have to tailor things your own way to optimize performance. But, here's my list of 14 extra tips to consider, provided alphabetically, not in order of importance.

Buy a calendar. Purchase a physical paper calendar that has hourly or daily time slots that you can use to schedule your priority tasks. Or, if you prefer, use the hourly and daily calendar functions that automatically come with your smartphone, smartwatch, tablet, or laptop computer.

Delegate and outsource minor tasks. Once you've determined which of your many tasks are truly important, your time and energy should really be spent on them. But, even unimportant tasks may need to be done eventually by someone. If possible, delegate these to someone else in the workplace, outsource them to other departments in your company if they have more expertise, or even outsource them to other businesses who specialize in those particular tasks.

Draw a flowchart. It can be very helpful to visually see the task-by-task workflow of a project from start to finish. This can also be useful to help you create a milestone-by-milestone plan for your long-term career goals. Since many engineers are very visually oriented, this can actually be more effective than simply writing down the same thing in words. After all, as the old adage says, "A picture is worth a thousand words."

Get an accountability partner. Find a trusted family member, friend, or work colleague who will keep you accountable in your efforts at personal organization by regularly asking you if you've scheduled this week's tasks into your calendar, or if you've taken a few hours this month to review your long-term career goals. Maybe they can do this by telephone, by email, or over a weekly coffee break. And you can also do the same thing for them to keep them accountable.

Have a planning time. Block off, say, an hour each week to review your prior week's progress on your various tasks. You can also use this time to prioritize and schedule all short-term tasks into your calendar for the upcoming week. Moreover, you can also block off a few hours each month or every several months to review and plan your long-term career goals. Otherwise, your career will be pulled in all directions.

Just say "no." It really is a simple, but effective, word. Peer pressure, fun trivial hobbies, unproductive personal habits, and even workaholism can pull us to say "yes," when we know we should really say "no" in order to focus on achieving our real career goals. But, the more we are able to discipline ourselves and remind ourselves of what we truly want, it becomes easier to just say "no."

Know your energy cycles. As the saying goes, some people are early birds (i.e., they like to get up early in the morning to start work), whereas other people are night hawks (i.e., they like to stay up late at night to focus on work). But, other people function most efficiently in the middle of the day. Your most energetic time each day depends on numerous physiological and lifestyle factors, so figure out what works best for you and see if you can schedule some of your most important tasks during those high-energy times.

Learn from the experts. Read a book, watch a video, listen to an audio, or attend an in-person or online seminar on time and priority management created by experts in this area who do this full-time for a living. And, you may be able to find an expert who can mentor you as part of a small group or one-to-one for a period of time. This can definitely save you time, stress, and resources.

Organize your workspace. Take a look at your desk, office, workshop, or lab. Is it organized in the most efficient and effective way? Does it invite or discourage unnecessary interruptions by others? Does it allow you to physically access files, books, phones, and other resources easily and quickly? Does it help you focus on what needs to be done, or are there too many distractions like noises, smells, and entertainments? Does it help, or hinder, you to have a small fridge to quickly have some food and drink? Does it help, or hinder, you to have a small couch to take a quick nap from time to time? etc.

Overestimate the time-on-task. Often things take a lot longer than we originally thought. Maybe we thought we could finish a task in 1 hour, but it took 2 hours. Maybe the meeting was only supposed to last 45 minutes, although it ran for 75 minutes. Maybe we purchased something from a supplier who promised to deliver the item to us in 1 week, but it took 3 weeks. So, when we're scheduling daily tasks into our calendar and estimating timelines for completing projects, we should allow for a margin of error.

Schedule your breaks. Sometimes, we can get so extremely busy with work, that we may not even have time to take a break to eat, rest, socialize, or do nothing at all. If this is a perpetual problem, then it might be helpful to start scheduling break times into your daily and weekly calendar. But, if you do this, the break time needs to be treated as equally important as the work time. It gives you a chance to mentally and physically recharge, so you can be more efficient when you get back to work.

Stack your time blocks. In your daily calendar, one trick for your meetings with people is to schedule them into time blocks that occur one right after the other. That will often help ensure that meetings don't run overtime. Telling people you have to go to another meeting is also a legitimate excuse not to engage in pointless small-talk with people who like to linger after meetings.

Use the 80/20 rule. Have you heard about the 80/20 rule, also called Pareto's Principle? Based on good evidence, it states that about 80% of your successful results come from about 20% of the things you spend your time, effort, money, and resources on. Don't get stuck on the numbers. The reality could be anything like, say, 70/30, 84/16, or 92/8. The point is that you can prioritize your tasks by asking which ones will get you closest to your goal in the shortest time and with the smallest effort.

Write a vision statement. Crafting a vision statement for a project can help clarify and prioritize the daily tasks you have to accomplish for your short-term job duties. Just as importantly, a vision statement can help you determine long-term career goals, which might be different than what you originally thought. Tape these vision statements on the wall, save them to your computer screensaver, and so on. As the old proverb says, "Vision without action is a dream, but action without vision is a nightmare."

My Journey Toward Personal Organization

Looking back on my engineering education and career, I really wish that someone had shown me some techniques for managing my time and priorities. But, like many folks, I just learned things as I went along, either when I really needed to, or sometimes by sheer accident. In any case, I'd like to share briefly the story of how I learned various lessons about personal organization skills and how they helped me in my engineering education and career.

In high school, I had a daily routine. After classes were done for the day, I'd come home, watch some television, eat dinner, and then go to my room to do homework for 3 or 4 hours. No one told me to do this, not even my parents. But, I guess it seemed obvious to me that, if I wanted to do well in school, this is what I needed to do. And, as a result, I did very well. At the end of high school, I graduated with the second-highest grades in my class and even won a few awards. In general, I didn't have too many other activities in life or school, so I didn't feel the need to learn about time and priority management.

But, there was one lesson I did learn in high school from another student about personal organization that helped me perform better. As far as I recall, that student was visiting my country on a student exchange program. He did particularly well in mathematics tests and exams, like algebra, calculus, and relations and functions. I asked him how he was able to achieve his excellent results. He told me he would create a study schedule about a week before an upcoming mathematics test or exam, after which each day he would practice solving a certain number of mathematics problems. It was a simple and common-sense idea. And that's what I started to do. I was less stressed, better prepared, and my grades improved.

Ironically, when I started to study for my bachelor of engineering (BE) degree at university, some strange "switch" suddenly turned off in my head. I began to socialize a lot more—but nothing negative or self-destructive—and I got involved with a student organization, so I focused on those activities. As a result, I started to procrastinate on completing assignments and preparing for tests, so I often wouldn't begin studying until very late at night, just before I needed to go to sleep. I didn't even wear a watch, I didn't pay close attention to course schedules, and so I was often late attending course lectures. And, at the end of the second year of my BE studies, my course grades were quite poor.

In the third year of my BE studies, I met with a friend of mine who ran the student organization I was involved with. He asked how my studies were going and how my course grades were. I told him I wasn't doing so well. He then asked if I thought I could do better, and I told him "yes." He then shared a compelling story with me about a friend of his who completed a master's of engineering (ME) degree and how that opened up all sorts of life and career opportunities for him. My friend encouraged me not to let my time at university go to waste and to think about my life and career with a long-term view. Suddenly, in that short 1-hour conversation, another "switch" turned on in my head. I realized I wanted to succeed! And so, for the remaining 2 years of my BE degree, I worked hard, prioritized my studies over some other social activities, and achieved excellent grades. I graduated successfully with my BE degree and even obtained a summer research scholarship from the government to work in one of my professor's research labs.

Next, I successfully applied and enrolled for an ME program at the same university. This required me to complete 4 courses and a major research thesis project. However, I was finding things a bit stressful because of the advanced nature of the courses and the experimental nature of my thesis project. One day, when I was meeting with another engineering student in the cafeteria, I noticed he had an interesting-looking book with him. It was a daily calendar or organizer into which he wrote down all his activities into time slots, like his course lectures, assignment deadlines, social activities, and so on. I don't think I had ever seen such a book before that moment. I was so intrigued by this daily calendar that I went to the university bookstore, purchase one for myself, and started to organize my time and tasks. This really helped me feel less stressed since I now had better control of my ME studies and the rest of my life.

Also during my ME studies, I read a book on time management. The message I got from the book—rightly or wrongly—was that I needed to be self-disciplined. I thought I needed to wake up 1 or 2 hours before everyone else to get a head start on my activities before other things distracted me. And, if I really tried, I could complete all my personal and professional activities, if I just scheduled them properly. So, I tried this for a while in combination with my new daily calendar technique. This didn't work for me too well because I realized I still had too many things I wanted to accomplish each day. I still felt a lot of stress and found myself running around doing all sorts of activities from morning until night. Part of my problem at that time was that I hadn't learned anything about prioritizing my activities and just saying "no." I thought I could do it all. But, alas, I couldn't. As the old saying goes, "There just aren't enough hours in the day."

I then moved to another university to start my doctor of philosophy (PhD) degree in engineering. This required me to complete 5 courses and a major research thesis project. After that, I started a 2-year post-doctoral fellowship (PDF) in engineering at the same institution. In the midst of these scholarly pursuits, I had a nice social life and was also involved in giving some leadership to a student organization with its various on-campus activities. I seemed to be managing things well. Then, one of the other leaders of the student organization from another campus recommended a book on personal organization skills. This was very pivotal for me. This book talked about priority management, rather than time management. The core of the book's message was that we each need to decide what our main goal in life is and, then, prioritize all our tasks around that goal. It gave me permission to just say "no" to all the trivial and minor tasks that often distracted me. So, I spent focused time trying to figure out what my main goal was, I wrote it down in a vision statement, and I proceeded to prioritize my activities to be in-line with that goal and eliminate, as much as possible, anything else. I also spent time at the start of each school semester planning and prioritizing my activities. Mind you, I didn't do this perfectly, but it was a turning point in my approach. This gave me clarity of thinking on personal organization skills that I didn't have before.

A few years after finishing my PhD and PDF, I moved to another city and eventually got a full-time permanent job as a junior engineer in charge of a hospital-based biomedical engineering research lab that was affiliated with a local university. It was during this time that I read a number of books on leadership skills, biographies of famous leaders, and the 80/20 principle. These gave me insights into how to further develop my ability to manage myself, but also how to manage others effectively, as more and more engineering and medical students became involved with my research lab. Also, I continued to practice just saying "no" when I diplomatically, but clearly, asked my boss if I could just focus my research activities on biomedical engineering, rather than the clinical patient studies he was starting to involve me with. After all, engineering was what I was really qualified for and, thus, I felt I could do a better job doing that, rather than getting into an area that was really not my interest or expertise. Fortunately, my boss agreed. That allowed me to really focus my daily activities and boosted my overall productivity in completing projects and getting the results published as articles in peer-reviewed academic journals.

After several years, I settled in long-term as a senior engineer and research professor. I continued to learn about personal organization skills through experience and various resources. I tried to implement them rigorously. So, I became highly focused and selective about which book writing or editing invitations I accepted, which scientific reviewer tasks I completed for journals, which scientific conferences I attended, and which ME or PhD students I supervised, rather than saying "yes" to everything. I drew flowcharts and used checklists to help me visualize the various personnel and projects I managed, rather than letting ideas just roll around aimlessly in my head. I updated my resumé every time I published a journal article or conference paper or when I graduated a new student, rather than procrastinating for months, only to be stressed out when I needed to dig up that information to provide my department with a progress report. I even eliminated some personal hobbies and habits—sometimes permanently, sometimes temporarily—so I could have more time and energy to focus on my primary career goals. As a result, I had much more satisfaction and success in my engineering career and in my personal life.

So, What's the "Take Home" Message of This Letter?

The essence of what I've written here is that deliberately organizing your priorities and time will make you more efficient and effective as an engineer. It can even make your career and life more enjoyable. Of course, I'm not an expert on this topic. I don't pretend to be. But, I have definitely invested enough time, effort, and resources to improve my personal organization skills to see real results in my own career. And, I'd encourage you to do the same. You don't need to learn these lessons the hard way like I did by trial-and-error. Rather, please allow me to suggest that you read just one book or watch just one video on the subject. Then maybe attend a seminar and perhaps eventually find an expert mentor who can coach you for a while on this important career skill. There's no telling what benefits you'll experience in your engineering career and your life too.

All for now,

R.Z.

Letter 17 Communication Skills for Engineers: Speaking and Writing

Dear Nick and Natasha,

I'm glad to keep communicating with you about engineering. So, in this letter, I want to deal with that very issue—communication! There's a negative stereotype about engineers, namely, that they can't articulate their ideas in a clear or exciting way. Now, it's true that many engineers and many people, in general, don't relish speaking to groups or writing documents. Because engineers need to communicate with each other, the boss, the client, the customer, etc., they should improve their communication skills. Sadly, most engineers simply "pick it up as they go" when it comes to this topic, but have never learned about it deliberately through a course, book, video, seminar, or expert mentor. And so, I'd like to discuss the issue.

Some of the ideas I want to share with you here may be new to you, while others may be self-evident; the trick is to put them into action! I find it personally challenging too. Also, many of the skills here are probably transferable to other fields of endeavor, like education, entertainment, family, politics, religion, sports, etc. Thus, I will give several practical real-life examples from either my own or a colleague's engineering career to illustrate how these ideas can be applied by engineers at work.

Moreover, maybe there are some questions that I don't fully address in this letter; even so, I hope it does motivate you to take a strengths-and-weaknesses inventory about your own communication skills and then take concrete steps to improve. Please note that I will use the word "company" to refer to any university, industry, or government for which an engineer might work. There's an old saying I once heard that says something like, "Reading gives you breadth, writing makes you precise, and speaking keeps you ready." With that, let's turn to our theme of communication skills (see Figure 17.1).

Speaking Skills

We all know what speaking is, don't we? It's the act of opening our mouths and making intelligible sounds called "words" in order to relay information to others. Yet, if we really want to be effective at it, the way we do this and what information we convey, in fact, really depends on the context. Don't underestimate the potential positive (or negative) impact of good (or bad) speaking skills on your engineering

DOI: 10.1201/9781003193081-17

Conference Presentations
'Free flow' Conversations
Job Interviews
Media Interviews
Progress Reports
Training Seminars
Thesis Presentations

Digital Communiques
Grant Applications
Journal Articles
Media Releases
Memos and Letters
Patents
Progress Reports
Technical Manuals
Textbooks
Thesis Dissertations

FIGURE 17.1 Communication skills for engineers often involve speaking (*top*) and writing (*bottom*).

career. So, here are some common types of formal and informal speaking engagements that engineers are often expected to do, general tips and tricks that you can consider, and a few examples.

Types of Speaking

Conference presentations are talks that you may need to give at academic, industry, and government conferences. The goal may be to give an update on the results of your research on a new product or technique that you have been developing for your employer. Or, it may be to promote the benefits of the new commercially available product that your company now offers. These can be short or long, but should be formal in tone.

"Free flow" conversations are probably the most frequent verbal communication that you will have with your boss, client, customer, supplier, team, etc. These informal ordinary chats can occur over a coffee, in the office, on the shop floor, while

traveling, or around the meeting table. They should be clear, friendly, purposeful, and respectful, but some silence or bluntness may be needed at times. These can be short or long and may be formal or informal in tone. And sometimes you might find out some very important information that would otherwise remain hidden from you in a formal meeting.

Job interviews that you participate in could be for a new position or promotion inside your own company. Please take these as seriously as the interview(s) that led to your job with your employer in the first place. Or, perhaps you might be secretly contacted by an employment recruiter hired by another company to lure talented engineers away to themselves. These can be short or long, but are usually formal in tone.

Media interviews that you might engage in could be for radio, television, newspaper, magazine, internet blog, internet podcast, etc. The media source that interviews you could be independent, run by the government, or even owned by your own employer in order to provide internal updates to everyone inside your company, to post on your employer's website, or to give as press releases to the public. These can be short or long and may be formal or informal in tone.

Progress reports are presentations that you might need to give to your boss, client, customer, team, etc., in order to update them on the current status of the product or project you are working on. This could be composed of a brief verbal description, a lengthy discussion, an in-person tour or demonstration, an audiovisual presentation, or some combination of these things over the better part of a day. These may be formal or informal in tone.

Thesis presentations are talks that you have to give to your university supervisory committee to complete a master's or doctoral degree in engineering. Many companies will financially and morally support you to do this, while still employed with them. They will get a much more highly skilled worker who further advances their goals, while you could get a salary increase and promotion. These can be short or long, but should be formal in tone.

Training seminars are guided instructional events that you may need to facilitate for your boss, client, customer, team, etc. You might be mentoring them on the use of a new computer software package, a piece of machinery, an electronic device, and so on. This can be a short event that lasts, say, an hour, or it may be a long event that lasts, say, a day or two. These are usually formal in tone.

Tips and Tricks for Speaking

First, know your audience. Who will you be speaking to? What do or don't they already know? What do or don't they want to hear? What language barriers or cultural customs should you accommodate? etc.

Second, know the format. Will it be formal or informal in tone? What are the time limits? Where is it going to be done? What kind of physical space will you be in? What audiovisual resources will be available? Will there be a time for questions and answers after your talk? Will it be permitted to digitally record what you say or is this talk off-the-record? What is the dress code? etc.

Third, know yourself. Do you speak clearly or mumble? Do you need to have some drinking water nearby to keep your voice clear? Do you tend to overuse a certain word or phrase? Do you habitually make certain body gestures that are helpful or annoying? Do you ever say inappropriate things, like jokes or criticisms, which make others feel uncomfortable? etc.

Fourth, practice your talk. Is there time to rehearse your talk adequately? Are there any friends or colleagues that can hear your talk and give honest and useful feedback? etc.

Fifth, get some help. Have you ever tried resources to help improve your formal and conversational speaking skills, like a mentor, book, video, seminar, or organization? If not, why not? etc.

Example 1: Effective Speaking

I'd like to demonstrate how giving interviews to the media, if done well, can benefit your company and you. I know an engineer who worked for a company that was developing a new product that could solve a simple, but very widespread, societal problem and could result in massive product sales to the public.

That engineer gave several television interviews to well-known media outlets with the full support of their employer because my colleague was the real expert on the topic within the company. Initially, my colleague had concerns about having a good physical appearance, the challenge of talking to multiple cameras during the interviews, the media possibly misquoting or misrepresenting the product concept, and the potential of competing companies stealing the idea. Fortunately, the interviews went smoothly.

The benefit to my colleague was that these interviews were prized additions to their resumé, which somewhat helped them to secure a university professorship later on, which really was their career goal. And the company benefitted because the interviews greatly increased the number of visits to their website by a factor of 5, and then several other companies saw the interviews and partnered with my colleague's employer to fabricate the product and bring it to the marketplace.

Example 2: Effective Speaking

Here, I'll illustrate the power of leveraging the ordinary "free flow" conversation. I once built up a friendly, but professional, rapport by email and telephone with another engineer in charge of product manufacturing for a certain company, and I often bought supplies from them for my biomedical engineering research lab.

Then, that engineer happened to be visiting my area, so we arranged for our first in-person meeting at a coffee shop, which went very well. We followed this up later by arranging to meet at an academic conference, along with another engineering professor I knew from a local university. The 3 of us discussed some potential research projects we could collaborate on that would benefit us all.

The result of these informal exchanges was that the company gave us an official contract and funds for a research project to evaluate one of their new products, we published several peer-reviewed academic journal articles and conference papers based on our analysis of that product, and we obtained "preferred customer" status that gave us a cost discount whenever we purchased that product.

Writing Skills

We all know what writing is, don't we? It's the act of putting on paper or inputting in digital format a series of symbols called "words" so we can convey information to others. But, if we truly wish to be effective at this, how we do this and what information we share, in fact, is really influenced by context. Let's not underestimate the possible benefits (or drawbacks) of strong (or weak) writing skills on someone's engineering career. So, here are some common types of formal and informal writing tasks that engineers are often expected to do, general tips and tricks that you can incorporate, and a few examples.

Types of Writing

Digital communiques, like emails, social media posts, and cell phone texts, are probably the most common kinds of writing which you will utilize as an engineer in the workforce. These are usually short and can be formal or informal in style.

Grant applications are written proposals to funding agencies run by private entities, government bodies, industry departments, or professional engineering societies to obtain funds or other support for your company, department, or team. These can be short or long, but are always formal in style.

Journal articles are scholarly articles in which you describe your cutting-edge research results. The goal is to have them published in recognized journals. This is the most common way for your new research findings to be communicated to the broader academic or industrial community. These can be short or long, but are always formal in style.

Media releases are nontechnical mini progress reports you might write on the development, performance, availability, and sales of a new product you or your team, department, or company have developed. These can be posted on your employer's website, social media sites, etc. These are usually short, but may be formal or informal in style.

Memos and letters, in paper format or by email, are used internally within your company, department, or team, and are usually not meant to be shared publicly. They may deal with a social topic, policy issue, or technological problem. These are usually short and can be formal or informal in style.

Patents are scientific and technical documents in which you describe the basic operating principles for a new product that you or your team, department, and company have invented. A patent will prevent other competitors from stealing your idea and making a financial profit at your expense. These can be short or long, but are always formal in style.

Progress reports are documents you write to give to your boss, client, customer, team, etc., in order to provide them with an update on the current status of the product or project you are working on. These can be short or long, but are always formal in style.

Technical manuals, sometimes called product brochures, are documents that provide instructions on how to assemble, operate, maintain, repair, and/or dispose of a product that your company has produced. They may only be for internal use within your company, or they may be included with the product when the customer makes the purchase. They can be short or long, but are always formal in style.

Textbooks are scholarly technical books written by experts for students or other experts on a particular specialty of engineering or a certain emerging technology. If you or your team, department, or company have become recognized experts in this regard, then producing such work can help you shape the field, as well as greatly benefiting your peers in their work. These books are long and formal in style.

Thesis dissertations are scholarly documents in which you describe in detail the background, methodology, and results of a major research project to complete your master's or doctoral degree in engineering. Many firms will financially and morally support you to get these degrees, while still working for them. They get a much more highly skilled worker, while you probably get a salary increase and promotion. These dissertations are usually long and are always formal in style.

Tips and Tricks for Writing

First, know your audience. Who will you be writing to? What information do or don't they already have? What kind of material do or don't they want to read? What language obstacles or cultural niceties should you consider? etc.

Second, know the format. Should you be formal or informal in tone? Is this meant to be a private or public communication? Is there a certain structure or template that must be followed? Are there word or page limits? Is there a deadline that needs to be met or can your communication be sent anytime? Are you permitted to keep copies for your company's or your own records? etc.

Third, know yourself. Do you write clearly and concisely or do you tend to get sidetracked? Do you often overuse certain words or phrases? Do you ever write inappropriate things, like jokes or criticisms, that make others feel uncomfortable? Are your spelling and grammar good? etc.

Fourth, practice your writing. Do you have time to write and rewrite several drafts? Do you have friends or colleagues that can read your material and give honest and useful comments? etc.

Fifth, get some help. Have you ever used any resources to help improve your formal and informal writing skills, like a mentor, book, video, seminar, or organization? If not, why not? etc.

Example 1: Effective Writing

I'd like to highlight the fact that an engineer—despite the negative stereotype that engineers can't communicate effectively—can indeed use their writing skills to produce textbooks that educate and influence generations of engineers and engineering students in university, industry, and government. After all, who else is going to write educational tools for engineers?

Dr. Russell Charles (R.C.) Hibbeler received his bachelor's degree in civil engineering, a master's degree in nuclear engineering, and a doctoral degree in theoretical/applied mechanics. He's worked in industry and also as a university professor. With all this experience, he's written a series of university-level textbooks that have been used by many people, including myself, over the decades and that have been reprinted many times. Just a few of the titles include *Structural Analysis, Mechanics of Materials*, and *Engineering Mechanics: Statics and Dynamics*.

Similarly, Mr. Tyler G. Hicks completed his bachelor's degree in mechanical engineering. He's worked in industry, taught at a number of engineering schools, and traveled widely to give lectures around the world. He's also brought his skill, experience, and knowledge to become the prolific author or editor of over 100 books in engineering and in related areas. Among the many works he's written or edited is a series of handbooks with titles like *Civil Engineering Formulas, Handbook of Chemical Engineering Calculations, Handbook of Civil Engineering Calculations, Handbook of Energy Engineering Calculations, Handbook of Mechanical Engineering Calculations*, and *Standard Handbook of Consulting Engineering Practice*.

Example 2: Effective Writing

I want to describe the typical process of writing and publishing an academic journal article in the field of engineering. For my team and me, our first step, of course, is to complete a mechanical/biomedical engineering research project. Then, we find a suitable journal whose publishing scope coincides with our research topic and then write our article to meet the journal's formatting requirements, which typically includes these items in this order: title page, abstract, introduction, methodology, results, discussion, conclusions, references, figures, and tables. Next, we circulate the first draft of our article to everyone on our team to incorporate any final corrections, after which one of us acts as the corresponding author who uploads our article to the journal website.

Then, the journal arranges to have our article reviewed by anonymous expert peer-reviewers in the field who provide feedback on our article and either invite us to revise it or reject it. Now, if we are invited to revise it, we make all appropriate changes to it, provide a point-by-point rebuttal letter to the reviewers, and resubmit it to the same journal for potential acceptance; but, if our original or revised article is rejected outright, we then submit it to another journal. This iterative process continues until, hopefully, the article is officially accepted for publication in a journal, after which we receive from the journal a final "proof" copy of our article to double-check for typos.

Note that, after we initially submit our article to a journal, it may take from several months to a year or so for it to appear online or in-print in its final published form. Thankfully, I've been reasonably successful in publishing many research articles, some of which have been quite popular and made a useful contribution to my field of mechanical/biomedical engineering.

So, What's the "Take Home" Message of This Letter?

The key thing I wanted to bring to your attention is the importance of speaking and writing skills for an engineer. It's something I need to be reminded of too. If we can learn how to effectively communicate, it really can help us achieve our professional goals. Sometimes effective communication is the only difference between you and your greatest competitor, whether that's to get a particular job, be promoted to a new position, attract a new customer, or be awarded that grant to fund your project. Although a lot of what I've discussed may be obvious to you on a theoretical level, the challenge is to put it into action in a concrete way. And so, allow me to encourage all of us to avail ourselves of the many good resources out there to improve our communication skills as engineers, not just for our own sake, but also for the betterment of society at large.

Seize the day,

R.Z.

Letter 18 Leadership Skills for Engineers: How to Lead Others Effectively

Dear Natasha and Nick,

I'm pleased to keep interacting with you about engineering. I've been somewhat busy these days, but you probably have been too. As you know, a core part of engineering is about designing, building, inspecting, repairing, or disposing of a device, structure, software, database, or process. Sometimes these tasks can be done by yourself. But, often you will work as part of a team, and sometimes you actually may lead a team. Add to that the individual interactions you might have with your supervisors and those working under your supervision, and you start to realize there's a lot more to engineering than merely technical skills. In these situations, an engineer needs to know how to lead others effectively and also how to be led.

Unfortunately, most engineers just "learn as they go" when it comes to this issue, but have never taken a course, read a book, watched a video, gone to a seminar, or been mentored by an expert. And so, I'd like to focus on this topic here. Thus, the purpose of this letter is to define leadership and then highlight 12 leadership skills. Although many of the skills are obvious, the trick is to put them into practice. To be sure, many of the skills are applicable to any endeavor, whether it's art, business, education, family, religion, politics, sports, and so on. So, I'd like to give a real-world example from my own engineering career for each leadership skill, so you can see how these ideas can be applied by an engineer; but, I know I still have so much to learn about being a good leader.

Also, I'm sure you can think of other leadership principles, rearrange my ideas in different ways, and find all sorts of resources from real experts who study and teach leadership development on a full-time basis. Nevertheless, I hope this letter at least inspires you to find out more, so you'll continue to develop your own leadership skills as an engineer. With that said, let's turn to our topic.

What Is a Leader?

A leader can be defined as someone who has certain skills that help them mobilize people and resources to achieve a particular aim (see Figure 18.1). A leader could be highly experienced or just starting out, aggressive or humble in personality, and the smartest person in the room or of just average intelligence. The people and resources could be vast or few in number. The aim could be grand or modest in

DOI: 10.1201/9781003193081-18

FIGURE 18.1   Leadership skills for engineers.

scope. But, in all cases, a leader is a person who achieves tangible results or at least tries their best to do so. And, whether they realize it or not, a leader will inevitably influence people around them for better or for worse. As an old saying declares, "More is informally caught, than is formally taught."

Leaders Have Vision

A leader has a picture or image in mind of the ultimate future outcome of anything from a small project to a large endeavor. This will help the leader know exactly what they want to achieve and how to communicate that to others. As the old proverb says, "Vision without action is a dream, but action without vision is a nightmare." It is very useful to carefully create a concise written vision statement and even practice saying it out loud verbally.

For example, I once had an idea for developing a series of new biomedical implants. To help focus this broad idea for myself, I crafted a written vision statement something like this: "My vision is to establish a biomedical engineering research program dedicated to the design, fabrication, and validation of in-novative, patentable, and marketable spine and bone fracture implants made from advanced new materials that solve real-world medical problems." It's short, simple, and clear.

Leaders Make Plans

A leader implements a deliberate step-by-step process, whereby the vision for anything from a small project to a large endeavor can be achieved. As the old saying goes, "If you fail to plan, you are planning to fail." The plan must describe what specific tasks and goals will be completed and what budgets, personnel, supplies, timeframes, etc., will be required. Depending on circumstances, the plan may be developed by the leader alone or with their team, but it can also be used to effectively persuade others to buy into the idea too. Sometimes, the greatest challenge for a leader making plans is to differentiate between a good plan, a better plan, and the best possible plan.

To illustrate the point, let me turn back to the same biomedical implant idea I mentioned earlier. I spent months preparing a long, detailed, written, research plan, as well as giving a series of talks to hospital, university, industry, and government stakeholders. The result was that 3 of the 4 stakeholders bought into my plan and agreed to participate in its realization. The 1 stakeholder that didn't join up did so due to an internal policy they had, but not because they didn't agree with the scientific or technological merits of my plan. Although this could have been potentially frustrating for me after all the work I did, I quickly came to realize that this is a common outcome in the world of engineering. Not everybody is going to agree with you. So, the other 3 stakeholders and I joined together to take the next steps to put my plan into action.

Leaders Recruit Teams

A leader finds or invites a group of people to use their own unique experience, knowledge, and skills in order to work together to achieve a common goal. Certainly, an individual leader can make a big difference, but quite often a team is able to accomplish far more. Sometimes a leader will inherit a pre-assigned team, which gives the leader no control over an individual member's pre-existing strengths and weaknesses. Or, a leader will need to recruit people to the team, which gives the leader a little more control over its final composition. Regardless of the different personalities, knowledge, experience, and skill among the team members, a good leader will try to accomplish a few key things. In particular, an effective leader should work hard to organize the team, function as an integral member of the team, delegate tasks, resolve conflicts, and encourage the team to take ownership of the goal, rather than seeing and treating the team as a mere instrument of the leader's will. As the old adage says, "There's no I in TEAM."

In my case, I know that I personally haven't had all the experience, knowledge, and skills needed to complete many biomedical engineering research projects on my own. So, I've routinely recruited multi-disciplinary teams of mechanical engineers, materials engineers, machinists, technicians, and orthopedic surgeons who have a variety of expertise in computer modeling and analysis, experimental testing, fabrication techniques, and surgical procedures. One of my main tasks in these team contexts has always been to create an environment where everyone truly feels that their contribution is immensely valuable and that everyone has the opportunity to fully make that contribution. Moreover, whenever I have more flexibility in choosing the people I recruit, I look for people whom I consider to be FAST. This stands for "faithful" (i.e., they have a track record of doing what they promise), "available" (i.e., they have the time to fully participate in the project), "skilled" (i.e., they have prior skill or knowledge in the area of need), and "teachable" (i.e., they have an open attitude to learning new things).

Leaders Train People

A leader invests the necessary time, effort, and resources to pass on to others their own skill and knowledge. This benefits the leader by reducing the pressure on them, so they don't have to micromanage the activities of individuals and teams under their supervision, but it also builds up their own resumé with additional leadership experience to make them more desirable in the workforce. This benefits individual trainees by building up their resumés with more skill and knowledge to make them more marketable in the workforce, but it also benefits individuals and teams to better accomplish stated goals.

And, of course, organizations and companies benefit from quality training of employees because it grows the roster of capable workers within their ranks. At times, there is a fixed formal training process that an organization or company uses that the leader simply implements among the trainees. At other times, the training process can be more flexible, informal, and tailor-made depending on the previous skill and knowledge of the trainees or the specific needs of the organization or company at a given moment. As the old proverb asserts, "Give a person a fish and they will eat a meal, but teach a person how to catch fish and they will eat for a lifetime."

Now, I have trained many engineers under my supervision on all kinds of testing equipment and fabrication machinery using an acronym that I've developed called SHOW. I always start with S which stands for "show"; this means I personally perform the task once or several times while the trainee simply watches what I do. Then, I move on to H which stands for "help"; this means I ask the trainee to perform the task once or several times while I simply help them do it. Next, I move on to O which stands for "observe"; this means I ask the trainee to perform the task on their own once or several times while I simply observe and make helpful comments in case there is a problem. Finally, I move on to W which stands for "walk away"; this means I physically leave the area, room, or building while the trainee continues to independently work on the task until the project is complete, but of course I make myself available to the trainee to deal with any questions or problems that may arise.

Leaders Mobilize Resources

A leader obtains any physical supplies that are required to reach the leader's aims, such as computers, equipment, materials, physical space, software, stationery supplies, tools, etc., but finances are certainly going to be a factor also. A leader should think through what resources are needed, what quantity of each resource is needed, whether these resources need to be procured permanently or temporarily, how much the resources will cost, how to obtain funding if needed, and so on. But a leader also has to be realistic about what resources they already have and what new ones can actually be obtained; thus, they may need to adjust their vision, even if it will result in more modest accomplishments. As the old saying goes, "A bird in the hand is worth two in the bush."

For example, early in my career as a junior engineer, I became the day-to-day manager for a hospital-based biomedical engineering research lab. I soon realized that our research projects could be completed much more effectively without relying as much on outside machine shops for fabrication and outside instrumentation for testing, if we invested additional lab funds to purchase more supplies. So, I approached my supervisor with the idea, got his approval, and proceeded to purchase all sorts of new manufacturing equipment (e.g., band saws, circle saws, drill presses, grinders, etc.), mechanical testing instrumentation (e.g., strain gage system, thermographic camera, etc.), and a large assortment of new hand tools and power tools (e.g., digital caliper, digital screwdriver, etc.).

Leaders Achieve Win-Win Goals

A leader wants every individual or group involved in a project or endeavor to get some tangible return benefit from their investment of time, effort, resources, skills, etc. The type of benefit may have been agreed to before the project began and can vary widely depending on what is ethically and legally appropriate: it could be financial payment, renewal of a work contract, improved reputation in the field, co-authorship on publications, and so on. As the old adage says, "Every bird wants to wet its beak a little in the water."

At other times, however, perhaps the individual or the group does not even want any return benefit because, as part of their own decision or policy, they have the mandate to make donations or do charitable work as a way to generously give back to others and society at large. In the normal course of events, though, leaders should try to achieve win-win goals for everybody involved.

Let me tell you how a colleague (whose engineering team was university-based) and I (whose engineering and medical team were hospital-based) collaborated a number of times on biomedical engineering research projects. Their team did the computational modeling, while my team did the experimental testing. Their team paid the salaries, while my team purchased the supplies. Their team wrote a first draft of the research article, while my team edited it thoroughly and submitted it to a peer-reviewed journal for publication. And so, it was a fruitful win-win partnership.

Leaders Ask Good Questions

A leader asks insightful questions of others, as well as of themselves. Such questions can cause us to think more deeply and creatively about our aims, plans, tasks, etc., so they can be carried out more effectively. Such questions may even fundamentally challenge the basic underlying assumptions of a system, organization, or institution. At times, a leader's questions may be welcome because there is a willingness to make the changes necessary to improve things. At other times, a leader's questions will be resisted because there is a desire to maintain the status quo for personal or ideological reasons. Whatever the case, leaders need to decide when it is appropriate to ask good questions and when it is better to remain silent. As the old proverb says, "Choose your battles wisely."

As a case in point, my role as a research professor of mechanical/biomedical engineering has been to supervise students in mechanical engineering, materials engineering, orthopedic surgery, and the like. And one of the primary goals of this is to conduct research that is unique and that will make a new contribution to the scientific literature for the betterment of the biomedical engineering field itself, the healthcare system, the medical professionals, the patients, and also society at large. If there is nothing new in what we are doing, why are we doing it? There is no point in reinventing the wheel. Unfortunately, it is easy for anyone to get caught up in the details of the task at hand without asking whether it is worth doing or not. So, one of my habits to refocus my team's research has been to continually ask my students and myself this simple, but useful, question: What new contribution will this research project make?

Leaders Support People's Ideas

A leader will encourage other people's ideas, interests, and goals because it may greatly benefit the leader, team, organization, or company in unforeseen ways. A good leader realizes there may be a lot of untapped creativity and potential in people that should be stirred up, rather than being crushed because the leader is afraid to lose control, inflexible in their methods, or short-sighted in their plans. This takes a lot of psychological maturity and social skills on the part of the leader because people's egos, emotions, and hopes are involved when they bring new ideas. Leaders who encourage people will earn a lot of respect. As the old saying goes, "People don't care how much you know, until they know how much you care."

Obviously, if a person's ideas are directly opposed to, or could cause great harm to, the leader, team, organization, or company, then it may be best for that person to be released so they can go in their own direction; but, this should be done as amicably as possible. Such a separation can actually be good because it can reduce everybody's stress and allow them to pursue their own agendas in an unrestricted and productive way.

For example, I have often made sure to ask engineering and medical students if they think there is a way to perform a stress test, compute a value, fabricate a specimen, or perform a surgical procedure that is even better than I originally suggested to them; this simple inquiry sometimes generates a better solution than I originally envisioned. Similarly, I have often told engineering and medical students that I am always happy to write them a good recommendation letter for any new position they apply for, if they decide not to continue with my team after their contract or studies are completed; it's amazing how such simple acts of respect have helped grow my network of contacts and professional collaborations with a number of these former students and employees.

Leaders Give Credit

A leader should privately and publicly reward and recognize other people's strengths, contributions, and achievements. This is fair and generous to the other folks, it will motivate them to stay productive, and it will earn the leader respect. All of this can create a positive atmosphere in the team, organization, or company.

In contrast, a leader who hordes all the accolades by claiming or letting others think that they alone accomplished everything will foster resentment and breed enemies. No one wants to put in a lot of time, effort, and resources just so somebody else gets rewarded or recognized. No one wants to be a mere stepping stone on the roadway to someone else's career success. And so, a leader should understand and respect the fact that other people have their own career aspirations and encourage them in that regard. As the old adage says, "Give credit where credit is due."

To illustrate the point, I've always ensured that everyone on the teams I've led received proper credit as co-authors on the biomedical, mechanical, or materials engineering research articles I've published in peer-reviewed scholarly journals. Sometimes it is challenging to determine who on the team should have the prized position of being the first author. But, at other times it is very clear which person has done the brunt of the hands-on work and, so, they are acknowledged as the first author; after all, they deserve it.

Leaders Are Reliable

A leader must turn their promises into actions. Others inside and outside the team, organization, or company should see the leader as someone who fulfills their promises to produce tangible results. As the old proverb says, "Say what you mean, and mean what you say." This kind of reliability gives others a sense of security because they know what to expect from the leader. Of course, leaders are not perfect, so they will sometimes overpromise and underdeliver. But, if they acknowledge their flaws and people see them making improvements, then their reputation for reliability can be restored.

Conversely, if the leader always blames others for failures and is unwilling to make improvements, this can ruin their reputation for reliability. Then, others will think the leader is incompetent or even dishonest. That can create disappointment, frustration, and stress, as well as reducing productivity. Although this kind of negative pressure can still push people to produce results quickly to meet upcoming deadlines, in the long-term it can lower the quality and quantity of the work being done.

Now, an important task in my engineering career has been to complete research projects and make sure our team's efforts result in articles that are published in peer-reviewed academic journals. Due to my own mistakes in managing time, personnel, or resources, there have been some projects I didn't complete, but fortunately, this has rarely happened. The vast majority of the time, however, I have earned a reputation for fulfilling my promises and producing concrete results. And so, some of my colleagues have given me nicknames like "The Finisher" and "The Manuscript Monster."

Leaders Know How to Follow

A leader needs to have prior experience being supervised or mentored as an individual or as part of a team. And a leader may also currently be under the supervision or mentorship of another leader that has a more senior position. Either way, this will help a leader empathize with the questions and challenges of those they are leading. It will continue to give them formal and informal training in developing their own leadership abilities. And it will keep them accountable to others for their actions.

In contrast, a leader who has never been a follower, who thinks they have nothing to learn from others, and who believes they don't have to answer to anyone for their actions is on their way to being a petty tyrant who misuses their position for personal ambition. This person has disqualified themselves from true leadership. As the old saying states, "Absolute power corrupts absolutely."

Personally, I have worked on biomedical, mechanical, or materials engineering projects as an ordinary member of a team, or in a supporting role as a co-supervisor, under the leadership of more junior colleagues. I have tried to encourage their growth and respect them in their leadership, even if I had more years of experience on the project's topic. And so, I made sure to say things to them like "You're the boss on this project" or "Whatever decision you make is okay with me" or "That's just my suggestion, but it's up to you, if you want to do it."

Leaders Know When to Quit

A leader knows when it's time to let go of something and move on to the next thing. There is often a natural growth-and-decay process that people, projects, and organizations go through. There may be an initial rise in enthusiasm, funding, recruitment, or productivity, followed by reaching a steady-state or plateau, and then an eventual decline. In some cases, the process may be quite normal and expected, so there is no need for a leader to control or worry about it. This can occur when an employee's contract comes to an end, the project reaches completion, the funding goal is met, the machine has reached target efficiency, and so on.

However, in other cases, the process may be negatively affected by internal or external factors, so it becomes destructive to participants and resources. This can happen if a person is consistently not performing their duties or meeting deadlines, the project is way over budget, the machine continues to produce too many flawed products, and so forth. At some point, a leader may need to step in to restart, redirect, or end things. In some ways, these kinds of decisions may be extremely difficult for a leader's ego or career, since it implies that they have failed in some way. Yet, the decision to end something or change course may, in fact, be the wisest thing a leader can do. Also, there's no shame even in losing or letting something go. And it's good for our own personal character development to be humbled once in a while and be reminded of our own flaws and limitations. As the old adage declares, "It's better to retreat from the battle now and then come back to fight another day."

As a case in point, I've had to quit or experience failure a number of times in my own engineering career. I've had to give up on publishing some research articles and books because I did not have the time or energy, although the opportunities were there. I've had to stop applying for certain grants to obtain project funding because my proposals were repeatedly rejected by decision-making committees. I've had to accept the fact that I wasn't hired for potential jobs that I was really excited about. And, sadly, I've had to dismiss some well-meaning engineers from their jobs or tasks because they were consistently not meeting the performance expectations that were previously agreed to.

So, What's the "Take Home" Message of This Letter?

The main idea in this letter is that you too can continue to expand your leadership skills as an engineer, as opportunities arise. I don't think there is such a thing as a perfect leader; that shouldn't stop you from attempting your best. And I definitely have much to learn and apply in my own leadership development. Although leadership can certainly be greatly enhanced by experience, personality, and intelligence, an equally important factor is simply the willingness to try and to grow. So, let's get out there, find some good books, seminars, websites, and videos, and then move forward in becoming the best leaders we can possibly be.

Good luck,

R.Z.

Letter 19 From Screws to Space Stations: All Engineering is Important

Dear Nick and Natasha,

Here I sit at my computer keyboard in a bit of a pensive mood as I begin to write to you. One thing that has always bothered me, and perhaps you too, is that sometimes engineers question the validity of their careers. After all, what real contribution have they made to human civilization? What device, machine, system, software, or process have they invented that has been truly remarkable? I know I've had these kinds of discouraging doubts about my own engineering career from time to time.

On the whole, of course, the engineering profession has made major contributions to society throughout the ages. I don't think I have to remind you of all the technologies that have sprung from the minds and hands of those in our noble line of work. And we have even had geniuses emerge from our midst who are now household names and whose work continues to shape our world. And I take pleasure in that knowledge. But, I am really talking about the individual engineer who sometimes feels their particular work is a bit dull, a tad unimportant, and maybe even altogether pointless. Yet, why should that be so?

On the contrary, it seems to me that all engineering is important. None of it should be taken for granted. If we take a step back and take a look at what engineering has produced, we'll see several lessons. First, most advances in technology are usually evolutionary, not revolutionary; most occur gradually and build on the work that came before. Second, ideas and inventions that are seemingly trivial and those that are apparently crucial, often have to practically work together hand-in-hand in order to achieve the desired goal. Third, some engineering benefits society in useful everyday ways, while other engineering benefits society by inspiring a big vision; they are both equally needed for human society to keep moving forward.

It's one thing to say all this and another thing to really know it, feel it, and act on it. So, in this letter, I want to encourage and even persuade you that the engineering that you do—whatever it might be—is really vital. To illustrate this, I want to describe 2 engineering marvels that seem to be on opposite ends of the spectrum of importance and excitement, but which have changed the world in their own unique ways. I beg your indulgence if, at times, I seem to be stating or describing the obvious, yet I do so to make a point. Here are the stories of the screw and the space station (see Figure 19.1).

DOI: 10.1201/9781003193081-19

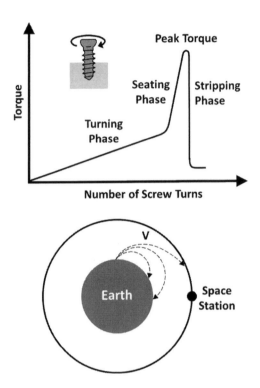

FIGURE 19.1 The screw (*top*) and space station (*bottom*). *V* is the escape or orbital velocity.

The Story of the Screw

Historians of technology tell us that the screw was invented around the year 350 BC probably by the ancient Greek engineer Archytas of Taras who, incidentally, was a friend of the philosopher Plato. Screws were not that commonly used by the ancients because the external (or male) threads had to be cut and filed by hand. Yet, by the 1st century AD, the engineer Heron of Alexandria described a screw thread cutting device that could make internal (or female) threads.

But, what exactly, is the screw? The screw is a device made from metal, plastic, or other material, whereby an angled plane is wrapped helically around an axis to produce external teeth-like threads. The screw head, however, has unique cuts, slots, or other features that can receive the tips of tools called screwdrivers. Screws are made using various fabrication techniques, such as 3D printing, lathes, and dies.

How does the screw work? Essentially, when an optimal torque is applied via a screwdriver to the screw head, it advances the screw tip forward into 2 or more materials in order to fasten them together with an optimal holding force. This happens in the following way.

In the initial *turning phase*, only a small rise in torque is required to turn the screw around its axis and advance the screw farther into the materials. This torque is caused by the friction generated as the screw threads cut into the materials to be joined in order to create matching internal thread-like grooves.

In the *seating phase*, a rapid increase in torque is required to turn the screw, which ultimately results in a peak torque and peak holding force. These peak levels are caused by the screw head fully being seated or pushed against the top of the materials to be fastened.

However, if the screw is turned beyond this point, then the *stripping phase* occurs. When this happens, the external threads of the screw and/or the internal thread-like grooves of the materials begin to strip or break. This, in turn, causes a rapid decrease in the torque and holding force.

And we engineers are always glad when there's a formula that describes a phenomenon. In this case, it is $T_{\text{PEAK}} = 0.5\, F_{\text{PEAK}} d\, (p + fd)/(d - fp)$, where T_{PEAK} is the peak torque, F_{PEAK} is the peak holding force, d is the screw outer diameter minus $0.65p$, f is the friction factor between the screw material and the object material, and p is the screw thread pitch which is the distance between 2 neighboring screw threads.

Now, the bolt is also a type of screw with external threads, except that it passes freely through a smooth hole in the materials to be fastened, and then is tightened into a nut with matching internal threads. A machine screw is also a type of screw, except that its external threads are tightened into pre-existing matching internal threads in the materials to be fastened. The same formula mentioned earlier applies here.

The screw is almost everywhere we look today. It holds together many items that people use every day, like appliances, containers, cupboards, doors, electronics, furniture, plumbing, shelves, vehicles, and so forth. It also fastens together parts used to make special commercial, industrial, and scientific equipment, like airplanes, bridges, computers, machines, medical implants, microscopes, telescopes, trucks, satellites, and so on. A large airplane, for example, can use several million screws and other fasteners.

It's no exaggeration to say that, in the history of the world, many more screws have existed than people. There are probably trillions upon trillions of screws in use at this very moment as I write this letter. And when screws break, it is often merely annoying. But, it can also result in great damage, financial loss, and even human death. The poet-priest George Herbert wrote a 17th-century proverb that says, "For want of a nail the shoe is lost, for want of a shoe the horse is lost, for want of a horse the rider is lost." Similarly, I'd say that for want of screws or bolts our modern technologies and, ultimately, our civilization will be lost. The screw is a simple, but fascinating and important, device for the engineer and for the world!

The Story of the Space Station

The aim of the International Space Station (ISS) is to advance knowledge to prepare humanity for its possible future in space. Beginning in the 1990s, the construction of the ISS was finally completed in 2010. At a cost of 150 billion US dollars, it was the most expensive engineering project ever completed in the history of humanity.

An endeavor of this magnitude necessarily involved the cooperation, resources, and money of national space agencies from different countries.

Consisting of countless individual components, the ISS required the designing and building expertise of many scientists, engineers, and technicians. This included computer modeling, electronics, manufacturing, materials, mechanics, orbital physics, radiation physics, robotics, and so on. The ISS was not fabricated on Earth and then launched into space wholesale. Rather, individual components were taken into space on at least 42 trips by rockets and shuttles and then assembled there. The Space Shuttle, for instance, used to take 27,000 kg of payload into space in a single trip and then dock with the ISS.

According to NASA, the ISS weighs 420,000 kg, its capsule portion is 109 m long from end to end, its solar panel portion is 73 m long from end to end, and its internal surface area is 3252 m^2. It can accommodate 8 spaceships docked to it at once, as well as 20 different research payloads, on the outside. A 16.8-m long robotic arm on the outside is used to move modules, do experiments, and transport spacewalking astronauts and cosmonauts. It has 350,000 sensors to monitor the health and safety of the crew, 50 onboard computers to control everything, 12.9 km of electrical wiring, and 8 solar panel units that provide 75–90 kW of power. It has pressurized sections accessible by the crew, as well as unpressurized sections that are not. It has 6 bedrooms, 2 toilets, 1 gym, 1 viewing bay window, and other areas.

But, how does the ISS stay in orbit? We know from orbital physics that any object launched from the Earth must reach an escape velocity to overcome the planet's gravity in order to enter orbit at the same velocity. Otherwise, it just crashes back into the planet. Like all artificial near-Earth satellites, the ISS travels in a circular orbit at a velocity that is mathematically expressed as $V = \sqrt{GM/R}$, where V is the escape or orbital velocity, G is Newton's universal gravitational constant, M is the mass of the Earth, and R is the orbital radius. So, the farther out the desired orbital path is from the planet, the slower the escape velocity that's required, and the slower the orbital velocity that's maintained. For the ISS, the orbital altitude is 400 km above the Earth at an orbital velocity of 29,000 km/hour.

Astronauts and cosmonauts from a number of countries have performed over 2,700 research experiments to understand the effects of low gravity on the human body and other aspects of biology, chemistry, and physics. Those stationed on the ISS—which can accommodate 6 people at once—have to first undergo months and years of rigorous physical, psychological, and intellectual preparation on Earth. Despite that, they know they are putting themselves in harm's way. They are exposed to deadly radiation from cosmic rays and solar flares and sometimes need to move to better-protected sections of the ISS. And those who complete a 6-month tour of duty on the ISS lose about 1% of their bone mass each month. They also experience muscle atrophy, immune system depression, cardiovascular degeneration, and optic nerve damage. They often need 1 year of physical rehabilitation once they return to Earth to recover from the effects of low gravity, but some of the damage is apparently permanent. The cost of human understanding, it seems, is human risk.

The ISS is a gateway into deep space. Yet, even gateways need to be designed and built by someone. The scientists, engineers, and technicians who did so remind us that vision, knowledge, skill, and cooperation can be mixed together to produce astounding outcomes. But, equally importantly, the ISS reminds humanity about the challenges and opportunities that still await us out there among the stars!

Which Is More Important?

After having now read about the screw and the space station, which do you think has contributed more to human civilization? It's really not meant to be a trick question. It seems to me that both have been, and continue to be, highly valuable inventions. I personally don't see it as a competition. But, I can see arguments for either as more important for those who wish to pick one.

On one hand, the screw is much more practically beneficial to us on a day-to-day basis, since it quite literally keeps our physical belongings from falling apart; but, it doesn't inspire us to nobility and greatness. On the other hand, the space station is the exact opposite; it doesn't really benefit us in performing the daily routines of life, yet it holds within itself the promise of a bright future for humanity's destiny. In my view, they each have an important role. The screw reminds us of the practical and sensible aspects of engineering, while the space station reminds us of the poetry and vision of engineering.

So, What's the "Take Home" Message of This Letter?

The idea that I want to get across to myself, as much as to you, is that all engineering has its proper place. All engineering, whether it appears trivial or grand, can make a difference to people's individual lives and to society as a whole. We engineers should never feel apologetic, cynical, or discouraged about what we do on a daily basis in the workplace, whether it's at the university, in some industry, or with the government. Sometimes it may be exciting, sometimes it may seem dull. But, it always, always, always, makes a difference to somebody somewhere sometime. So, let's go forth and change the world one screw and one space station at a time.

With best wishes,
R.Z.

Letter 20 Have an Engineer's Eye: Watching for Future Technologies

Dear Natasha and Nick,

Here I am once again writing to you about engineering, as requested. I some-times think that we engineers can be thought of as worker bees toiling away at a task that confronts us in that moment. Yet, we may be oblivious to the fact that the very skills and knowledge we possess—and even the technology, technique, or process we may be involved in—is something we, in part, inherited from our predecessors.

This was ably stated by the renowned physicist Sir Isaac Newton (1642–1727) who once wrote, "If I have seen farther, it is by standing on the shoulders of giants." Recognizing our debt to the past is all well and good. But, notice that his statement also focuses on using what lies behind in order to better see what lies ahead. And that's what this letter is about, namely, the future!

Now, I'm not going to play the futurist by trying to predict what technologies will emerge in the near or far future. Instead, I'd like to do 4 things below. First, I'd like to offer some philosophical thoughts about the overall nature of scientific and technological progress. Next, I'd like to suggest a few practical reasons for watching and embracing emerging technologies. Then, I'd like to give a few ideas about the various resources that we can consult—and their pros and cons—to keep updated on new emerging technologies. And finally, I'd like to finish with a few real-life stories.

How Do Science and Technology Progress?

I'm not going to try to answer this question in a comprehensive way here; after all, I'm not a professional historian. And there are entire books dedicated to this topic. However, I think there are some issues that are worth highlighting since they seem to show up in history over and over again. In particular, I'd like to briefly discuss the role of anomalies, hybrids, and peripheral factors on the fuzzy progress of science and technology that leads to new emerging ideas and products (see Figure 20.1).

Thomas Kuhn (1922–1996) wrote a classic book in 1962 called *The Structure of Scientific Revolutions* that's been influential and controversial in understanding how science advances into the future. Prior to his book, philosophers of science taught that science always makes advances through a step-by-step linear process of making an hypothesis, performing a test, and revising the hypothesis based on test results,

DOI: 10.1201/9781003193081-20

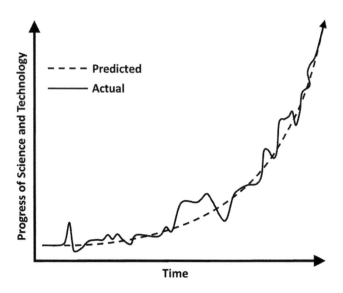

FIGURE 20.1   The progress of science and technology.

almost like solving a puzzle. But, Kuhn's analysis showed that the step-by-step process of normal everyday science sometimes uncovers *anomalies* that reveal flaws in old paradigms. These old paradigms eventually are unable to explain the growing evidence of anomalies and are forced to give way to completely new paradigms. Thus, he coined the term "paradigm shift." Science, therefore, progresses through long periods of evolutionary steps punctuated occasionally by revolutionary jumps. Of course, there are always science laggards who fight the new paradigm until they die or undergo a type of intellectual conversion to accept the new paradigm. So, it often takes a generation for the new paradigm to become fully established. Similarly, I would argue that technologies developed by engineers are accepted and used by society for long stretches of time, until new technologies suddenly emerge that prove to be superior in performance. These new products can then quickly replace the old products to become the new state-of-the-art.

In addition to anomalies, my opinion is that another factor that sometimes rapidly opens up new vistas for science and technology is *hybrids*. By this term I mean either temporarily bringing together interdisciplinary teams composed of different experts to achieve common goals, or permanently combining long-established specialties to create entirely new fields having their own experts. Without such short-term or long-term hybridization, these new areas of research and development would not be possible and, hence, would slow the progress of science and technology. And it is at these new intersection points between older fields of study that sometimes the most rapid progress is made in research and development. So, for example, mechanical engineers, materials engineers, and orthopedic surgeons often temporarily work together by lending their particular set of skills and knowledge to the team in order to develop novel medical implants for replacing bad joints (like hip and knee replacements) and repairing broken bones (like cables, nails, plates, and screws). Alternatively, the permanent merging of established fields like mechanical engineering, electronics engineering, and computer science has launched the field of robotics as a separate specialty and administrative department at some universities and technical schools. There are many other examples that could be given, but these are sufficient to illustrate the point.

And, finally, let's turn to what I like to call the *peripheral factors* (i.e., non-scientific and nontechnological) that affect the progress of science and technology either in a positive or negative way. This includes cultural trends, economic changes, epidemic and pandemic diseases, natural disasters, political upheavals, religious ideas, wars and rumors of wars, and so forth. These can cause jumps upward, drops downward, or plateauing in the advancement of science and technology. So, for instance, a change in political leadership in a nation can bring more or less government financial support to research and development depending on the promises, policies, and priorities of the new government. Similarly, epidemic and pandemic diseases can force universities, industries, and governments to funnel more money to scientists and engineers to rapidly find, produce, and distribute innovative cures, but diseases can also slow down progress if masses of scientists and engineers become ill or die. Or, military conflict can instigate a race between rival nations to see who can develop more advanced technologies to create more efficient weapons, but conflict can also destroy many valuable research facilities and kill many valuable researchers thereby slowing down science and technology. And religious belief can encourage science and technology if it teaches that God (or the gods) is a grand designer who created a logical and orderly universe that can be studied in a systematic way, but religion can discourage science and technology if it believes that God (or the gods) is fickle and erratic so that the universe has no regularity or laws that can be discovered.

Why Should Engineers Embrace the Future?

It's often said that the future is completely unknown. Although that's true in the strictest sense, it's possible to take a look at how the past has shaped the present, incorporate some lessons into our engineering work, and then take a reasonably educated guess as to how the present may potentially affect the future. One of the worst things we engineers can do is to be solely focused on solving the urgent problems of today without giving thought to how that might affect future generations in good and bad ways. I think there are excellent practical reasons for engineers to keep an eye on possible future technologies. Here are some realistic scenarios.

Let's say we are engineers teaching and training the next generation at *university*. We'll prepare them better for future careers in the workforce by informing them about new cutting-edge research that they may eventually encounter. Developing such quality students has a beneficial ripple effect. If our students go into the workforce better prepared in skill and knowledge, they will perform better in their jobs. This will, in turn, improve our university's reputation for engineering education, which means government, industry, and private donors will invest more money into our university. This will increase our university's ability to purchase or develop the best state-of-the-art teaching resources, which will attract even more students to our university to be educated as high-caliber engineers who contribute to society. And so on.

Or, for example, if we are engineers who are designing and building a product in *industry*, we can incorporate new emerging technologies to greatly improve the performance of our own product for our clients and customers. Developing such quality products has a beneficial ripple effect. If we give our clients and customers superior products, they will be happier because it will effectively solve their practical problems. Then, they will be more likely to hire us again to solve other problems, which will bring more money into our company. Next, this will also improve our company's reputation in society, which can bring in a whole new group of first-time customers and clients. The increased money from old and new clientele will strengthen our company's financial status, thereby securing our own personal jobs and allowing the company to better contribute its expertise to improve society. And so on.

Or, as another case in point, if we are engineers doing advanced research in a *government* facility, our knowledge of emerging technologies can improve our own research results and help us avoid the mistakes and dead ends of other researchers. Developing such quality research has a beneficial ripple effect. If we are able to conduct cutting-edge research that produces better findings, conserves resources, and saves money, then the government will be more confident in its research institutes and the research engineers who work there. This means more funding for the various research programs going on, thereby allowing us engineers to expand our research endeavors by purchasing more state-of-the-art equipment and taking on new projects. This, in turn, allows us to provide more and better solutions for various societal problems, making for a safer and healthier society that is able to be more economically productive.

And last, but not least, there is a *caution* we should take to heart, whether we are engineers in university, industry, or government. Although this is not the time for a discussion of engineering ethics, I want to briefly highlight one point—just because we can do something, doesn't mean we should. Power has to be tempered by wisdom. What I mean is that there have always been, and always will be, many emerging technologies for which we cannot fully see the future consequences, whether good or bad. So, just because something is a cutting-edge technology, doesn't necessarily mean we should wholeheartedly welcome it into our own engineering work. In some cases, we may not have much say in accepting or rejecting it, especially if we are only junior professors, junior engineers, or junior researchers. Our employers may be the decision-makers. But, let's at least be the kind of engineers who are consciously willing to evaluate the beneficial and destructive potential of any technology and then do what we can to accept or reject it.

Which Resources Can Keep Engineers Up-to-Date?

There are a number of ways engineers can keep updated on new scientific discoveries and emerging technologies and even be inspired to think of original ideas. These resources can help engineers, as well as their universities, industries, and governments, to enhance their own engineering endeavors. Keep in mind, however, that each of these resources has their own pros and cons, so it's probably a good idea to utilize several of these, rather than relying solely on one as your main source for information. Here are a few key suggestions, but maybe you can think of others.

Get to know a researcher. These scientists and engineers may work at a university, for a company, or in a government-run facility. They personally use analytical/mathematical modeling, computer analysis, experimental testing, on-site field observations, and/or back-of-the-envelope calculations to conduct their research. So, they are on the frontiers of discovery on a daily basis and would even be able to provide you with the most recent data that's been calculated or measured earlier the same day. However, these new findings may be hard for you to obtain for several reasons. Some researchers sign a non-disclosure agreement (NDA) with their employer or other sponsor so that technological secrets are protected until products are patented, reach the marketplace, and sold; there is a lot of money at stake. Also, many researchers don't want their methodologies or results to be made public prematurely, since their employment may depend on them being the first to publish new discoveries in peer-reviewed academic journals before their rivals do; there are careers, future funding, reputations, and awards at stake. And sometimes, in the case of governments and the military, the latest research and even the operating principles of older technologies are perpetually kept secret for national security reasons.

Go to research conferences. These face-to-face or online research meetings may be sponsored by scientific or professional engineering societies, universities, industries, and/or governments. They are primarily meant to give science and engineering researchers a chance to meet with one another to exchange ideas in private conversations, in public lectures, and using posters; specific meetings may or may not be open to non-members or non-presenters. Also, short research summaries called "abstracts" of individual projects are published online or collected into online or paper books called "transactions" or "proceedings"; specific items may or may not be accessible to the general public. This information, though, will eventually become outdated like any resource. Another thing to look out for at these conferences is industry exhibitors who wish to showcase their new or emerging products to conference attendees in the hopes of selling their goods to, and/or fostering potential partnerships with, researchers.

Go to fairs and competitions. Many universities and individual university departments sponsor science and technology fairs that encourage their researchers to give short presentations and create physical displays that highlight their latest findings. They do this to reach out to the general public, recruit new science and engineering students, garner the interest of potential private donors and industry sponsors, and allow researchers to interact with each other. Similarly, some companies and government agencies sponsor competitions open to any bright and enterprising individuals or teams from the public to come up with unique scientific and technological solutions that currently face the company or the society the government represents. The winners of these competitions often get some monetary rewards, potential job offers, and achievements they can proudly add to their resumé. The downsides of fairs and competitions are that they don't necessarily occur frequently in your specialty or close to you geographically, and there may be some proprietary information that you won't be allowed to obtain or use even if it is of interest.

Read research articles. These online or paper journals are the traditional and primary way that new results from science and engineering research are disseminated by experts for other experts. The articles that appear in these journals are almost always peer-reviewed by other experts in a single-blind or double-blind process, so you can be assured that the methodologies and findings have been scrutinized for quality prior to publication. The articles are scientifically and technically sophisticated and detailed, so they are excellent sources for your work. Journals in online and/or paper format can be purchased by individuals or institutions by subscription, which allows access to all content depending on the subscription level; alternatively, individual articles can be purchased online. Some journals, however, are deliberately offered by the publisher as "open access," which means anyone can go online, view, or download articles freely without limit. The major drawback of journal articles is that they are already out-of-date by the time they are published. The reason is that there is often a time lag of several months to a year or so between the time that authors initially submit their article to the journal to the actual time it is finally published.

Read media releases. These are short online or paper announcements made to the public by individual researchers, universities, companies, or governments about their latest scientific or engineering research results or product availability. If you search the internet, you can easily find such information on the individual's or organization's website or other social media posts. You can also often sign-up to automatically receive emails from the individual or organization about new developments. This information is eventually also conveyed to the general public via television, radio, newspapers, and so forth. Some of this information may inspire you and even trigger creative thinking for your own engineering work. The main disadvantage of these announcements is that they are often superficial from a scientific and technical point of view since they are really meant for the general public rather than experts in the field. Also, by the time a new product is announced to the public, it is likely already patented or patent-pending; thus, its basic design and operating principle cannot easily be stolen or incorporated into another engineer's work.

Read books and magazines. First, textbooks for universities that have recently been published contain mostly well-established core information that science and engineering students need to learn at university. In a sense, they are written by experts for emerging experts. But, textbooks may also have content about recent technologies, techniques, or processes provided in a sophisticated and detailed way that could directly benefit your engineering work. Second, books and magazines for the general public about science and technology can also contain well-established core information, but they can also have content about new developments that could, at the very least, introduce a new idea to you and inspire you to pursue a course of action. It's not likely, however, that the general public books and magazines will have enough scientific and technical content for you in the long term. Unfortunately, these types of literature go out-of-date very quickly and, thus, publishers often publish revised editions every so often.

Read and watch science fiction. On June 19th, 1988, the legendary science and sci-fi writer Isaac Asimov (1920–1992) was interviewed for the Public Broadcasting Service (PBS) television program *Open Mind.* Asimov stated that "Good science fiction tries to invent a society which is different from our own [society]—distinctly different—but which holds lessons for our own [society] … And a friend of mine says that science fiction writers are scouts sent out by mankind to survey the future." Now, we all know that many sci-fi books, short stories, television shows, internet shows, and movies have been produced over the years. But, did you know that many of these works were created by people actually trained as real scientists and engineers? Some sci-fi works are high-quality masterpieces that have inspired generations of storytellers, scientists, engineers, and society in general. Other sci-fi works are low-quality excursions meant only for superficial and temporary entertainment and are quickly forgotten. Some sci-fi stories envision a bright future where science and technology have solved many of the world's problems, whereas others see a darker vision in which the misuse of science and technology has brought the world to the brink of self-destruction. Whatever the case, sci-fi tales commonly explore ideas that are far ahead of their time by proposing future technologies and societies that the ordinary engineer maybe would never even imagine on their own. Other sci-fi tales explore ideas more solidly rooted in established principles of biology, chemistry, and physics. But, all these tales can inspire engineers to think outside-the-box and then use their skill and knowledge to push the limits of their work, thereby advancing human civilization into the future.

Watch educational television and videos. Science-oriented and technology-focused biographies, documentaries, and programs can keep you updated on the latest and greatest developments, as well as about problems and disasters. Most of these aren't developed by experts for other experts, but rather by storytellers and journalists who want to inform the general public. So, their scientific and technical content may not be sophisticated or detailed enough to be of any immediate use to you, but they may introduce you to some new concepts. Also, as with any resource, these can also go out-of-date very quickly.

A Few Real-Life Stories

The television series *Prophets of Science Fiction* profiles important science fiction writers who've predicted, or strongly encouraged, real developments in science and technology. One episode highlights Jules Verne's classic 1865 novel *From the Earth to the Moon.* The novel tells the tale of Americans who use a giant cannon to shoot a 3-man capsule from southern Florida to reach the moon in 3 days. The passengers experience weightlessness on their journey to the moon, and then they splash down in the ocean upon their return to Earth. Amazingly, this describes in exact detail the actual mission to the moon and back over a century later in 1969 with the Apollo 11 spaceship, except that a rocket instead of a cannon was used. Similarly, another episode discusses Sir Arthur C. Clarke's award-winning 1979 novel *The Fountains of Paradise.* The novel tells the story of an engineer who conceives of a tower that reaches 36,000 km straight up into space and acts like a space elevator for transporting people and goods to a spacecraft docked at the tower's top. Although originally proposed by a real-life Russian physicist in 1895, the idea was periodically revived, but never caught on seriously because of technical challenges. Clarke's novel went a long way in popularizing the idea again. Consequently, by 2011, there were at least 60 universities in the United States, as well as NASA, seriously trying to turn the idea into reality.

This next vignette demonstrates how keeping an eye out for educational television and videos about cutting-edge research—although mainly meant for the general public—can bring about powerful benefits to various stakeholders. An engineer I know once worked for a company that was doing research and development of a brand new product that could solve a basic, but very widespread, societal problem. If successful, the product could result in massive sales to the public. This engineer gave several television interviews to well-known media outlets with the complete support of their company because this engineer was the real expert on the topic. The interviews went very smoothly, and they were broadcast in due course. The engineer's company benefitted immensely because the interviews greatly increased the number of visits to their website by a factor of 5. But, perhaps even more importantly, there were several other organizations who saw and were impressed by the television interviews and eventually partnered with the engineer's company to fabricate the new product and bring it to the marketplace.

The following brief personal anecdote shows that published research articles are a good way to find out about the latest findings in a particular field. There are more impressive examples I could give, but I retell my own experience to highlight that most research like my own is a slow step-by-step process that typically moves science and technology forward in small ways. Now, my own research as a mechanical engineer has been focused on biomedical applications. You could say that I'm really a biomedical engineer. My work has involved measuring the mechanical properties of biological tissues, doing computational and experimental stress analysis of medical implants, and developing new biomaterials for those medical implants. I've had the good fortune to often publish my new findings as articles in peer-reviewed scholarly journals. Other engineers, scientists, and surgeons have kept track of some, but not all, of my newest work by the moderate number of private requests I've had from peers to provide them with copies of a few of my articles, the high volume of visits and downloads a few of my online articles have achieved, and the frequent number of times a few of my articles have been cited by other researchers in their published works.

So, What's the "Take Home" Message of This Letter?

My core point has been that we engineers should keep a watchful eye on new scientific findings and emerging technologies. The reason is that it can inspire us with innovative ideas and help us make important improvements in our own work. So, if we work at the university, are we willing to discover and learn how to use emerging new tools for teaching our engineering students? Or, if we work in industry, are we willing to discover and learn how to use emerging new tools for designing, building, inspecting, repairing, or disposing of our products? Or, if we work in government, are we willing to discover and learn how to use emerging new tools for conducting our research? I would encourage us not to be too reluctant in giving up old tried-and-true ways to embrace the future unless there are good practical or ethical reasons for doing so. I hope this has been a useful and up-lifting read.

Best regards,
R.Z.

Letter 21 Protect Your Intellectual Property: Patents, Copyrights, and More

Dear Nick and Natasha,

I hope this letter is a pleasant surprise to you. I sometimes have to hastily—but, I still hope, thoughtfully—write these letters to you in-between my other duties. And maybe you'd say the same thing about reading my letters. In any case, whether we are engineers working in a university, in industry, or with a government, our work will deal with new ideas and products. These may have originated from our own minds and hands, or those of others. But, they are all extremely valuable. Consequently, a new idea or product is often called intellectual property (IP).

Sadly, we live in a world where IP can be misunderstood, stolen, ignored, lost, and even expire. Due to carelessness, ignorance, or downright malice, there are individuals and groups who can take advantage of you for their own benefit. So, it becomes important for engineers to understand how to legally obtain and protect their IP. Yet, many engineers are unaware of how to prevent themselves from becoming victims of breaches, or even how to prevent themselves from accidentally becoming the villains in the story.

That's what this letter is about—protecting your IP (see Figure 21.1). Keep in mind that each country has its own particular IP laws, which are often connected with other national IP laws through agreements or treaties; so, it's important to know the specifics. However, what I'm going to do below is provide general guidelines that do not depend on the IP of any particular nation. First, I'll discuss the main ways to become the legally recognized inventor and/or owner of your IP. Then, I'd like to describe why you should legally protect your IP. Next, I'll outline the different ways IP can be misunderstood or violated. And I'll finish with a few real-world engineering examples. In addition to reading this letter, I'd also encourage you to always seek advice from an educated and certified legal expert on these matters. Now, let's get to it.

How Can You Protect Your IP?

There are a few ways that allow you to become the legally recognized inventor and/ or owner of your IP. Some of these methods have greater legal strength than others, while others don't have any legal force but mainly rely on moral obligations

DOI: 10.1201/9781003193081-21

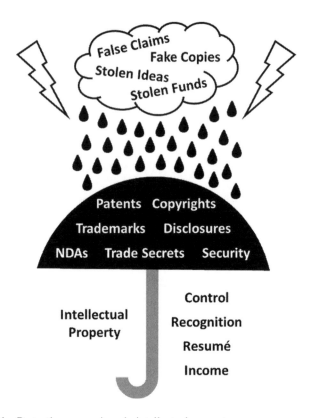

FIGURE 21.1 Protecting an engineer's intellectual property.

between participants. In some instances, you may want to keep your IP totally hidden from all public gaze, as we'll see.

A *patent* is a certificate from a national government that gives legal IP protection to an individual or group as the inventor and/or owner of an idea or product, such as a device, software, technique, or process. A patent owner has the legal and exclusive right to make, market, and/or sell the idea or product in that nation, while also preventing others from doing so. A patent owner may or may not temporarily lease their patent to another person or group for a one-time fee or ongoing royalty payment, so that another person or group can make, market, and/or sell the idea or product. A patent is usually granted for an idea or product that is new, useful, and not obvious to others in the field. A patent is generally not granted for discovering laws of nature, phenomena in nature, or mathematical formulas or equations. A patent usually costs money to obtain and maintain, and it may or may not expire by law after a number of years, after which the idea or product is in the public domain; then, anyone can make, market, and/or sell the idea or product. Depending on the nation, a patent could be issued based on the principle of the first to invent vs. the first to register vs. the first to use, so it's always a good idea to know your country's patent laws.

A *copyright* gives legal IP protection for a written or artistic work, such as an article, book, drawing, graph, photo, audio, video, or computer software. A copyright is usually automatically and legally assigned to the creator of that work when it comes into existence and/or when the creator writes the word "copyright" or enters the symbol "©" beside their name on a physical or digital copy of the work. A copyrighted work does not necessarily have to be registered with a national government or international registry, since it's automatically and legally recognized by many national and international laws. A copyright does not cost money to obtain or maintain. A copyright owner can choose to make, market, and/or sell the work on their own, while also preventing others from doing so. A copyright owner can also sign a contract to give a publisher either temporary or permanent and exclusive or non-exclusive permission to make, market, and/or sell the work, while the work's original creator obtains no payments (e.g., scholarly journal articles), a one-time payment (e.g., magazine articles and scholarly books), or ongoing royalty payments (e.g., scholarly books). Permission to copy or quote any part of the work, unless stated otherwise in the work, must always be obtained from the copyright owner, who may be either the creator or the publisher at a certain time. A work may potentially be in the public domain in a particular nation so anyone can copy, distribute, and/or sell it if the work's creator has specifically stated so, if the copyright protection has expired, and/or if the work is considered very old and was created before a certain year.

A *trademark* gives legal IP protection for a name, word, phrase, or symbol that is used to identify an individual, organization, or company and, hence, its products and services. A trademark needs to be registered with a national government to be legal. Then, no one else has the legal right to use or copy that trademark without expressly given permission from its owner. Also, the owner may or may not ask for a user fee from someone else who has gained permission to use the trademark. A trademarked name, word, phrase, or symbol is usually accompanied by "TM" (i.e., trademark), a circled "®" (i.e., registered), or something similar to let others know it is owned by someone. A trademark usually does not expire over time as long as it is being actively used by the owner.

A *disclosure* is a dated, signed, and witnessed pre-patent confidential statement that an idea or product, such as a device, software, technique, or process, was invented by a person or group. It is usually created if the financial cost of obtaining and maintaining a patent cannot be paid by the inventor due to their momentary personal or financial circumstances. It does not need to be registered or approved by a national government. But, a disclosure may or may not give legal IP protection in a particular nation. Even when a disclosure doesn't have legal status, it does have some moral authority. This means that professional peers and rivals who are aware of it will feel obligated to respect it, so they don't get a bad reputation in their field as an idea thief. Disclosures are not just useful before a patent is obtained, but they can potentially also be created before official copyrights and trademarks are secured.

A *non-disclosure agreement* (*NDA*) is a dated and signed agreement that the signatories will not privately or publicly reveal any details about an idea or product that is still being developed, such as a device, software, technique, or process. There are no financial costs involved, it does not need to be registered or approved by a national government, and it only expires under the conditions and by the date agreed upon in the NDA. Failure to comply could result in lawsuits, but only if the offended party has the money and means to pursue legal action against the person who broke the agreement. The reason is that an NDA may or may not automatically give legal IP protection in a particular nation. Nevertheless, it does have some moral authority because signatories will feel obligated to respect it, so they don't get a bad reputation in their field as being dishonest and untrustworthy.

A *trade secret* is a financially profitable and/or particularly critical idea and product, such as a device, software, technique, or process, whose details have deliberately been kept hidden by the inventors. The inventors don't want to get legal IP protection through a patent (or a copyright in the case of drawings, photos, or written works) because that would require revealing some technical details. In industry, there is a lot of money at stake, so the inventors don't want their rivals in the marketplace to make a similar competing idea or product. Also, in the military sphere, governments may not want to reveal certain technical details about weapons, security systems, strategies, policies, etc., so that unfriendly nations don't get hold of this information. The other benefit is that trade secrets never expire, unlike patents that may expire by law in some countries. Consequently, inventors put in time, effort, money, and resources for cybersecurity and facility security to keep secret the technical details of their idea or product.

Why Should You Legally Protect Your IP?

There are a number of very good reasons why you as an engineer should try to protect your IP through a patent, copyright, or trademark, unless you want or need to keep its details hidden as a trade secret.

Well, firstly, it lets you stay in *control*. If you are the legally recognized inventor and/or owner of your IP, it's quite natural to want to determine its fate. You can decide to make, market, and/or sell the new idea, product, work, etc., or not. No one else can tell you what to do with it. You are in charge.

Another benefit is professional *recognition*. Your reputation as an important and influential person in your field or industry can be boosted when word gets out that you are the official person who thought of and invented a brilliant new idea, product, work, etc. This, in turn, can open up other doors of opportunity for you professionally. And you may be able to leverage this to get a salary raise from your employer.

You can also add this item to your *resumé*. Your resumé will be greatly enhanced if you are able to state that you are the official IP owner of an idea, product, work, etc. This is a major accomplishment. You should be proud of it. And, if you decide to apply for a new job, this item will greatly impress the potential employer, since it says that you are a high-caliber candidate that stands out from other applicants.

It will give you another stream of *income*. If you or your company make, market, and/or sell your idea, product, work, etc., permanently to customers or allow customers to use it through user fees, then you can personally obtain some share of the profits. Or, you can temporarily lease your legally recognized IP to another person or group to give them permission to make, market, and/or sell the item, while you get a certain percentage of the profits (i.e., royalties). It's also possible for you to permanently sell your legally recognized IP to another person or group, which can potentially make you a lot of money.

How Can Your IP Be Violated?

There are several ways you as an engineer can be robbed of the recognition and rewards of being the inventor and/or owner of your IP.

First off, your IP *may not really be yours*. Let's say, in a very realistic scenario, that you put in a lot of effort to develop a new idea, product, work, brand name, etc., during your own private time at home without using your employer's resources. You might think that your new idea, product, work, etc., automatically and legally belongs to you, in other words, that it's your IP. Not true. It actually may belong to your employer, whether it's a university, company, or government. After all, they don't pay your salary just to be nice to you. So, long before you expend a lot of effort generating new concepts, check the details of your employment contract and related documents so you understand how much, if any, of the IP actually belongs to you while you work for your employer. Such policies can vary greatly from one employer to the next.

Also, your IP can be *stolen*. Let's suppose you had a casual conversation with a friend or colleague, or you posted something online to your friends on a social media website, about your new idea, product, work, brand name, etc. Or, let's say your computer, desk, or office were not properly secured. But, this was before you applied for and got legal recognition as the inventor and/or owner of your IP. Then, there is a potential risk that someone can copy your idea, product, work, etc., apply for and get full legal recognition for it themselves, and then they have legal IP status, not you. It may or may not even legally matter if you have documents, photos, videos, or eyewitnesses that say you developed the idea, product, work, etc. What matters is what the law recognizes, so you should be familiar with national and international laws for IP.

Moreover, your IP can be *ignored*. Let's say, in this case, that you are legally recognized by the law of one or more nations to be the inventor and/or owner of a new idea, product, work, brand name, etc. And perhaps you are even selling it in the marketplace. Some villainous people will ignore your legally recognized IP. They'll hack into your computer, have a spy working in your company, or break into your office, and then copy or download all the technical details of your idea, product, work, etc. Then, they'll manufacture clones of your gadget to sell it, or use your computer code to improve their own software, copy your manuscript, or just sell your technical details to others. Even more easily, they'll purchase your product legally, reverse engineer it to see how it works, and start making fraudulent clones they can sell. Depending on what country these scoundrels live in, or what legal marketplace or illegal underground network they use to sell their stolen wares, you may or may not be able to do anything about it at all. However, let's say that you don't have legal IP in another country and there's no patent treaty between your country and that other country. Then, those same people are not villains because they may possibly have the legal right to make, market, and sell your invention in that other country.

Additionally, your IP can be *lost*. Let's say, in another situation, that a student is asked to help with their professor's new idea. Or, perhaps an engineer is asked to develop their employer's new idea. The idea could potentially receive a patent, copyright, or trademark. The student and engineer might think that they have no potential IP ownership because it was not their original idea, although they are going to work on it in some fashion. Then, the professor or employer asks the student or engineer to sign a contract waiving any future IP rights. However, the student and engineer should both get legal advice before they sign such a contract. The reason is that those who conceive an idea like the professor or employer (i.e., conception) and those who bring it into reality like the student or engineer (i.e., execution) may both potentially have IP rights depending on the country. Of course, if the student and engineer are not interested in retaining any future IP rights, then they can go ahead and sign the contract.

Finally, your IP can *expire*. Let's consider the same case as discussed earlier, whereby you are legally recognized as the inventor and/or owner of the new idea, product, work, brand name, etc., in a particular nation or nations. The IP belongs to you as far as the law is concerned, but it can expire in one of two ways. In the first instance, you may need to pay fees on an ongoing basis to maintain your legal status as the IP owner, but if you default on these payments you could forfeit your IP ownership temporarily or permanently. In the second instance, your legal status as the IP owner may permanently expire by law after a certain maximum number of years, after which the technical details of your idea, product, work, etc., are then in the public domain. In both instances, this means anybody can now freely use your idea, product, work, etc., for personal or business purposes.

Example 1: Patent Wars

Nikola Tesla (1856–1943) was a famous electro-mechanical engineer biographized by Margaret Cheney and Robert Uth in their book *Tesla: Master of Lightning*. He had hundreds of inventions and patents to his name, including the brushless alternating current (AC) induction motor, fluorescent lights, the radio, radio remote control, wireless power transmission, the Tesla coil, the "Egg of Columbus," and so on. He received many awards and honors, for example, the strength of a magnetic field per square area is quantified by the scientific unit called the "Tesla" symbolized by the letter "T."

In 1887, Tesla applied for and eventually obtained 40 US patents related to his invention of an AC power system for producing and sending large amounts of electricity over long distances. Tesla then sold all 40 patents to the Westinghouse Corporation for a lump sum of cash, stocks in the company, and an ongoing royalty of $2.50 per horsepower for electricity sold to customers. But, by 1897, Tesla released the Westinghouse Corporation from royalty payments to ensure the company didn't go bankrupt because of its fierce "War of the Currents" rivalry with the direct current (DC) power system promoted by General Electric, the inventor Thomas Edison, and the financier J.P. Morgan. It was Tesla's AC, not Edison's DC, system that eventually won the "war" because it was technologically superior for powering cities.

In another situation involving patents, Tesla got embroiled in a controversy about the invention of the radio. In 1893, Tesla publicly demonstrated his system for sending, receiving, and tuning radio signals. In 1896, the inventor Guglielmo Marconi got a British patent for his system for sending and receiving a Morse code signal over 1.25 miles. In 1900, Tesla obtained 2 patents from the US patent office for his radio system. After being repeatedly rejected by the US patent office because Tesla and others had done similar earlier work, Marconi finally got a US patent for the invention of the radio in 1904, in part, due to the influence of big corporations. In 1911, Marconi won the Nobel Prize in physics, which irked Tesla so much that he sued Marconi's company in 1915 for patent infringement, although Tesla didn't have enough funds to pursue the lawsuit. Ironically, decades later just a few months after Tesla's death in 1943, the US Supreme Court re-examined Tesla's claims and patents and then officially named him as the primary inventor of radio.

Example 2: No Patent, No Reward

Sir Arthur C. Clarke (1917–2008) was one of the grandmasters of science fiction writing, which had earned him many awards and an international audience. His life and works are documented in detail by Neil McAleer in his biography *Sir Arthur C. Clarke: Odyssey of a Visionary*.

Interestingly, Clarke's university degree was not in literature, but in physics and mathematics. So, not surprisingly, he was a radar operator in the British air force during World War II. He did some experiments to see if he could bounce radio signals off the moon and receive them again, but it was too far away. However, he realized that if there was a manmade satellite orbiting the Earth that was much closer than the moon, radio signals could be bounced off the manmade satellite and received back on Earth.

These ideas eventually culminated in 1945, when Clarke wrote a seminal non-fiction article titled "Extra-terrestrial relays: can rocket stations give world-wide radio coverage?" in the magazine *Wireless World*. In it, Clarke suggested that 3 satellites equally spaced in a geosynchronous orbit—that is, rotating with our planet so they are always at the same location above the Earth—would be able to connect the entire world via radio signals.

A number of physicists and engineers read his article and then proceeded to make his concept a reality in 1963 by launching the world's first geosynchronous communication satellite into orbit at approximately 36,000 km above the Equator. This orbit is named in his honor as the "Clarke Orbit" or "Clarke Belt." The modern world couldn't function without this invention, since the modern cell phone, the radio, television, and other signals depend on these satellites.

The only problem, as Clarke later admitted, is that he didn't bother to patent the idea and, as a result, totally lost ownership and any financial benefit from his own invention. Later, he even wrote a 1960s essay about this incident, somewhat humorously titled "A short history of comsats, or: how I lost a billion dollars in my spare time."

Example 3: Benefits of Copyright

Researchers in all engineering specialties make their findings known to peers, students, and the broader field by publishing scholarly articles, book chapters, and/or books. So, I'll summarize my own experience as a published mechanical/biomedical engineer to illustrate what copyright is all about.

First off, my articles are published in peer-reviewed journals to whom I transfer copyright ownership. Then, the journals who own my articles benefit in these ways: (i) exclusive rights to be the first and only ones to publish and distribute my articles in all formats; (ii) their reputation grows as a source of scientific knowledge; (iii) they don't pay me anything; and (iv) they sell my articles for profit (i.e., traditional journals) or offer them freely online (i.e., "open access" journals).

But, I benefit too: (i) I don't need to spend any time, energy, or money to publish or distribute my articles; (ii) I can still use the content of my articles at scholarly meetings; (iii) I can still post my articles on my institutional website; (iv) I can still distribute my articles freely for private teaching and research; and (v) I can cite my articles in my resumé to prove to my employer and my peers that I'm a productive scholar.

Now, the book chapters I've written that have appeared in other people's books are produced by science and technology book publishers through the same process I just described above for my articles. As a chapter contributor, I receive authorship credit for my chapter as well as extra benefits like a small honorarium payment, a free copy of the book, and/or a price discount to buy additional copies.

Finally, my books as the sole author—or as the editor who recruits other authors to contribute their chapters—are also produced through the same process by science and technology book publishers. I get the same benefits as my articles and book chapters except that I can't post free copies online. But, I have the extra advantage of getting ongoing royalty payments from the publisher based on how many books are sold.

So, What's the "Take Home" Message of This Letter?

If you're an engineer with an original idea or product, you have created something valuable that can potentially benefit society. And, you deserve to be recognized and rewarded for your effort. So, protecting your IP only makes sense, as does respecting the IP of other engineers. I've also tried to offer a general, but useful, discussion on the ways and benefits of doing so. Remember to always seek advice from an educated and certified legal expert on these issues. I know that there is some time and cost involved in securing your IP, but if you are able to do so, it's worth it in the long run for you and for society.

All for now,

R.Z.

Part V

Aftermath

Letter 22 Engineering Your Retirement

Dear Natasha and Nick,

How is life treating you these days? For me, it sometimes seems to ebb and flow like a river. In any case, I'd like to tackle a subject in this letter that you may wish to consider seriously even right now—retirement! There's some evidence I remember reading somewhere (although I don't have the details at the moment) that people with no plan for active life during retirement don't do particularly well emotionally, mentally, or physically. Sometimes they can tend to deteriorate rather quickly.

And so, perhaps you already have notions about taking up a new hobby, learning a new language, traveling the world, volunteering with a charity, or just spending more time with family and friends; that's great! On the other hand, maybe you're very busy pursuing your career right now as an engineer and don't feel you have the time or energy or interest to start planning too far down the line for life after engineering; that's totally understandable!

Nevertheless, it might be beneficial just to be aware of what active engagement in science and technology can look like after formally retiring from engineering work. Some ideas I want to share, maybe things that you're already doing, so they won't be left to your retirement years. But, conversely, the extra time you'll have during retirement may give you a chance to pursue some of these options for the first time in your life. In any case, here are some ideas to consider plus a few real-world stories, if you're interested in something different than the standard retirement lifestyle (see Figure 22.1).

Active Engineering Society Member

You can be an official volunteer member of, and get actively involved with, a professional engineering society on local, regional, national, and international levels. The main function of these societies is to provide official certifications and accountability to their members, who are then recognized as practicing engineers by the government. But, they also create policies and standards for the engineering profession in general, have scholarship programs for engineering students, organize local activities for professional and student chapters, raise awareness among the public about the engineering profession, and so forth. There are plenty of ways to get involved.

Adjunct or Emeritus Professor

You can apply to a university engineering department to formally become a volunteer Adjunct Professor, which will give you an opportunity to supervise the

DOI: 10.1201/9781003193081-22

FIGURE 22.1 Retired engineers can still be active in science and technology.

thesis research projects of undergraduate and graduate students. You can then even publish articles about that research in peer-reviewed engineering and science journals. If you're already an engineering Professor, then consider staying on with an Emeritus status, which will provide the same opportunity to supervise and publish.

Ambassador-at-Large

You can volunteer as an engineering education ambassador for a university that will send you into high schools to speak about the engineering profession to inspire and recruit potential students. Similarly, some engineering firms could appoint you on a paid or unpaid part-time basis to be their ambassador among university students to recruit potential employees, as well as among the universities themselves to forge potential university/industry partnerships.

Board Member

You can be a volunteer board member of a charitable organization that promotes the legacy of an eminent scientist or engineer, lobbies the government to make policy changes for or against the use of certain technologies, etc., or, if you can't find a charity that's interesting, why not start one? Similarly, you can become a board member of an engineering firm that helps keep their organization economically efficient, environmentally friendly, and technologically productive. Almost always, these positions are unpaid, but it gives you a chance to shape the future, continue to build your resumé, and make potentially important personal contacts that could open up other doors of opportunity for you.

Engineer-at-Large

You can be an unpaid volunteer who helps your family, friends, charities, etc., to personally design, build, inspect, repair, or dispose of various household appliances or other devices, equipment, structures, toxic materials, and waste products. Most people are not very technically savvy and, once they find out you're a practicing or retired engineer, they are extremely grateful to receive your help in solving practical technological problems that to them seem insurmountable. And so, you can become an occasional engineer-at-large in your own social sphere of influence.

Freelance Consultant

You can start a small part-time engineering consultancy firm with just one employee—you! Certainly, you'd have to tap into your personal or professional network of contacts, advertise on the internet, place advertisements in engineering publications, etc., to get the word out about your services. This can bring in some needed supplemental income through consultation fees, which can be quite lucrative depending on the standard payment rates in your field.

Mentor-at-Large

You can make yourself available to communicate or meet informally with student engineers or working engineers to provide them with career advice and feedback. Perhaps these engineers in need of mentoring are people from your own family, friends that you know, friends of friends, or colleagues from your last workplace. You don't necessarily have to advertise yourself publicly as a mentor to engineers, but you can just be open to mentoring opportunities that arise spontaneously.

Scholarship Benefactor

You can personally donate or raise funds to establish an engineering scholarship fund at a university that would be awarded yearly to the most deserving engineering student. The scholarship could be based on course grades achieved by the student, a contest judged for the best speech or essay by a student, or some other criterion.

Many students struggle with finances to pay for good engineering education, so they rely on loans from family and friends, part-time jobs, etc., so creating a scholarship could be your way to give the next generation of engineers a chance to get started.

Technology Fair Participant

You can volunteer to be an organizer, judge, and/or keynote speaker for science and technology fairs and contests in which high school and university students display their knowledge, skill, and inventions. This will give you a chance to pass on some of your practical wisdom to the next generation and inspire them with a grander vision of how an engineer can contribute to society.

Web Show Host

You can start, produce, and host your own internet show on science and technology. It can be broadcast live or pre-recorded as an audio podcast or online video. With a minimal financial investment in some good quality software and equipment, this can all be done from the comfort of your own home. You can interview student engineers about their education and working engineers about their careers. You can answer questions sent to you by email or asked by live listeners and viewers. You can provide updates on the latest scientific and technological developments. You can do reviews of books and other resources. You can give practical tips and tricks about technical matters. You can share practical lessons you learned from your own engineering career. Although you can do this just as a fun hobby, it can also help mentor the next generation of engineers and even provide you with some financial income as a registered home business.

Writer or Author

You can write a personal online technology blog or even articles for newspapers and magazines on the latest technology news. You can write articles that provide state-of-the-art technical reviews on particular topics in your engineering field that can potentially be published in peer-reviewed engineering and science journals; this is a non-paid volunteer activity. If you're a little more ambitious, you could take on the daunting task of authoring a book on engineering, whether it is a university level textbook, a technical manual useful for industry, a journalistic-type book that tells the story of how a major engineering project was achieved, a "my life in engineering" autobiography, and so on; this, of course, could supply you with some added income through book royalties.

Your Final Contribution

You can state in your last will and testament that, after you leave this mortal existence, you want to donate some money to an engineering-related charity or project, an engineering scholarship or program at the university, an engineering firm, an up-and-coming engineer, and so on. Depending on your philosophical or religious views, you could consider donating your body to science, so that various scientists and biomedical engineers can conduct experiments that lead to a better understanding of the human body and that result in new technologies that improve healthcare.

A Few Shining Examples

I'd like to tell you briefly about a few officially retired engineers who are wonderful models of what I'm talking about.

One of my friends is a retired engineer, but he's made himself available as an engineer-at-large. He does this by sometimes visiting and working on his son's farm as an unpaid volunteer. He helps install, inspect, and repair various pieces of electrical and mechanical equipment. On one occasion, my friend called me by telephone to discuss some technical questions he had about a certain device. I'm afraid I wasn't much help to him, since I wasn't familiar with that particular apparatus. To some degree, my friend's efforts are not particularly surprising, since he cares about his son. Yet, the fact that he does this kind of work at all—rather than avoiding it because he's officially retired from engineering—is a testament to both his technical competence and moral character. And I admire him for it.

There's another retired engineer whom I met several times at various engineering events. He was a very active member of his professional engineering society (PES). This is the PES that gave him his official certifications and designations to work as a legally recognized professional engineer. During the time I knew him, he had several roles in his PES's local chapter. For instance, he was the secretary who kept track of nominations and volunteers for committee positions. He also made verbal announcements and gave updates at the monthly meetings that were open to members and inquirers. And he was on the professional review committee that interviewed and assessed those who applied for membership in the PES.

Another exemplar of active retirement is the visionary electro-mechanical engineer Nikola Tesla (1856–1943). In 1919, at the age of 63, he wrote a series of 6 autobiographical articles titled "My Inventions" for the magazine *Electrical Experimenter*, which also had Tesla's image on the cover. He described his early childhood, his first efforts at invention, his high school and polytechnic education in science and engineering, and his invention of some of his most famous electro-mechanical devices, like the alternating current (AC) induction motor, the Tesla Coil, and the ill-fated Tesla Tower. Although he was well past his creative peak when he wrote these articles, Tesla's name was still widely known in society at large. And so, the magazine editor hoped that Tesla's reputation could potentially sell magazines and even influence the next generation of scientists and engineers. The articles were eventually compiled into a book called *My Inventions: The Autobiography of Nikola Tesla*, which has been made available by various publishers over the years. Tesla kept himself occupied with engineering well into old age until his death. He continued to propose innovative devices to governments. He remained an official technical consultant with Westinghouse. He held yearly birthday dinners for the media to whom he would announce his latest technological ideas. And he still inspires many people today.

So, What's the "Take Home" Message of This Letter?

The main point I want to stress here is that your eventual retirement as an engineer does not necessarily mean that you can no longer have any meaningful involvement in science and technology. This involvement could be on a small personal level or a larger societal level. In your career, you will have accumulated experience, information, skill, and wisdom that you can pass on to the next builders of civilization. And, in the process, you can have fun doing it!

Peace and prosperity,

R.Z.

Bibliography

"Arthur C. Clarke", *Prophets of Science Fiction*. TV series, Discovery Communications, 2011.

"Asimov at 391", *Open Mind*. TV series, PBS (Public Broadcasting Service), June 19, 1988.

"Jules Verne", *Prophets of Science Fiction*. TV series, Discovery Communications, 2012.

Aristotle. *Mechanika* (or *Mechanical Problems*). Loeb Classical Library, www.LoebClassics.com.

Asimov, I. *Isaac Asimov: Visions of the Future*. Video, Analog SF Magazine and Quality Video, Inc., 1992.

Beveridge, W.I.B. *The Art of Scientific Investigation*. New York, NY, USA: Vintage Books, 1950/1957.

Capek, K. *R.U.R.* Prague, Czechoslovakia: Aventium, 1920.

Cheney, M., and R. Uth. *Tesla: Master of Lightning*. New York, NY, USA: Barnes & Noble Books, 1999.

Christensen, S.H., C. Didier, A. Jamison, M. Meganck, C. Mitcham, and B. Newberry, eds. *Engineering Identities, Epistemologies and Values*. New York, NY, USA: Springer, 2015.

Clarke, A.C. "A short history of comsats, or: how I lost a billion dollars in my spare time," in *Voices from the Sky*, edited by A.C. Clarke, 105. New York, NY, USA: Harper & Row, 1967.

Clarke, A.C. "Extra-terrestrial relays: can rocket stations give world-wide radio coverage?" *Wireless World*, October 1945.

Clarke, A.C. *The Fountains of Paradise*. London, UK: Victor Gollancz, 1979.

Copernicus, N. *On the Revolutions of the Celestial Spheres*. Schleswig-Holstein, Germany: Hansebooks, 2016 (original book published in 1543).

de Camp, L.S. *The Ancient Engineers*. New York, NY, USA: Dorset Press, 1960/1990.

Feynman, R. *The Pleasure of Finding Things Out*. Cambridge, MA, USA: Perseus Books, 1999.

Gugliotta, G. "One-Hour Brainstorming Gave Birth to Digital Imaging." *Wall Street Journal*, February 20, 2006.

Herbert, G. *Outlandish Proverbs*. London, UK: Humphrey Blunden, 1640 (Proverb #499 at www.GeorgeHerbert.org/Devotional/ProverbsTrans.html).

Heron of Byzantium (or Heron the Mechanic). *Heronis Mechanici Liber Des Machinis Bellicis*. Venice, Italy: Francesco Franceschi, 1572 (book title translated as *Heron the Mechanic's Book of Machines of War*).

Hibbeler, R.C. *Engineering Mechanics: Statics and Dynamics*. Upper Saddle River, NJ, USA: Pearson Prentice Hall, 2015.

Hibbeler, R.C. *Mechanics of Materials*. Upper Saddle River, NJ, USA: Pearson Prentice Hall, 2016.

Hibbeler, R.C. *Structural Analysis*, Upper Saddle River, NJ, USA: Pearson Prentice Hall, 2017.

Hicks, T.G. *Civil Engineering Formulas*. New York, NY, USA: McGraw-Hill, 2010.

Hicks, T.G. *Handbook of Civil Engineering Calculations*. New York, NY, USA: McGraw-Hill, 2016.

Hicks, T.G., ed., *Handbook of Energy Engineering Calculations*. New York, NY, USA: McGraw-Hill, 2012.

Hicks, T.G., ed. *Handbook of Mechanical Engineering Calculations.* New York, NY, USA: McGraw-Hill, 2006.

Hicks, T.G., and N.P. Chopey, eds. *Handbook of Chemical Engineering Calculations.* New York, NY, USA: McGraw-Hill, 2012.

Hicks, T.G., and J.F. Mueller, eds. *Standard Handbook of Consulting Engineering Practice.* New York, NY, USA: McGraw-Hill, 1996.

Kuhn, T.S. *The Structure of Scientific Revolutions.* Chicago, IL, USA: The University of Chicago Press, 1962.

Marcus Vitruvius. *De Architectura* (or *On Architecture*). Loeb Classical Library, www.LoebClassics.com.

McAleer, N. *Sir Arthur C. Clarke: Odyssey of a Visionary.* New York, NY, USA: Rosetta Books, 2013.

NASA (National Aeronautics and Space Administration), www.NASA.gov.

Newton, I. *The Principia.* Berkeley, CA, USA: University of California Press, 2016 (original book published in 1687).

Oxford 3000 Dictionary, www.OxfordLearnersDictionaries.com.

Star Trek, TV series, CBS Studios Inc., Hollywood, CA, USA, www.StarTrek.com.

Tesla, N. "Some Personal Recollections," *Scientific American*, June 5, 1915.

Tesla, N. *My Inventions: The Autobiography of Nikola Tesla.* Williston, VT, USA: Hart Brothers, 1919/1982.

The Big Bang Theory, TV series, Warner Bros. Entertainment Inc., Burbank, CA, USA, www.WarnerBros.com/TV/Big-Bang-Theory.

Verne, J. *From the Earth to the Moon.* Paris, France: Pierre-Jules Hetzel, 1865.

Wells, H.G. "The Land Ironclads," in *The Oxford Book of Science Fiction Stories*, edited by T. Shippey, 1–21. Oxford, UK: Oxford University Press, 2003.

Index

Printed and bound by CPI Group (UK) Ltd, Croydon, CR0 4YY

17/10/2024

01775690-0010